PRISTINE
SEAS

PRISTINE SEAS

Journeys to the Ocean's Last Wild Places

ENRIC SALA

FOREWORD BY LEONARDO DICAPRIO

NATIONAL GEOGRAPHIC

WASHINGTON, D.C.

CONTENTS

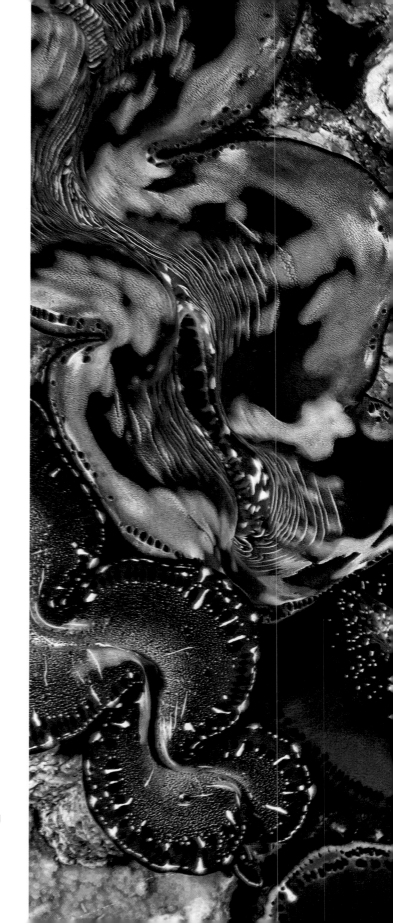

TROPICAL RICHNESS
A giant manta ray swims gracefully over a school of fusiliers in Palau. Both species feed on microscopic plankton that accumulates on the reef channel.

THE LANDSCAPE OF FEAR
Top predators rule in pristine seas—prey species are nowhere to be seen in the presence of abundant gray reef sharks in New Caledonia reefs. *(pages 2-3)*

HALLUCINOGENIC REEFS
Giant clams exhibit the most intense color palette in these seas. They pave the lagoons of pristine reefs like Kingman, in the northern Line Islands. *(pages 4-5)*

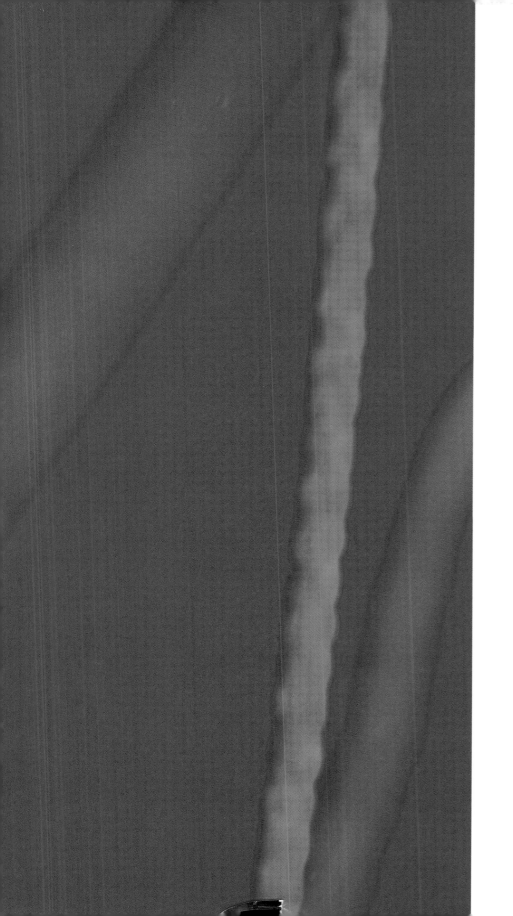

LIFE IN DISGUISE
A goby half the length of a fingernail lives among the branches of a sea fan in Palau's reefs, eating tiny invertebrates and hiding from predators.

FLOATING GOLD
In a saltwater lake known as Jellyfish Lake, found in Palau's Rock Islands, millions of golden jellyfish spend their days following the sun's movement across the sky.

LOVE DISPLAYS
Two male crosshatch triggerfish display
courtship postures in search of a receptive
female passing by.

Foreword by Leonardo DiCaprio

The world's oceans are a fascinating place. Their waters sustain a rich ecosystem that thousands of animal, plant, and fish species call home. I have always harbored a personal love for our oceans. As a child, long before I was ever an actor, I dreamed of becoming a marine biologist—and in junior high school, I made my first philanthropic contribution, donating five dollars to a save-the-manatee campaign in Florida. Despite never having seen a manatee before, I felt partially responsible for the plight of these animals that were dying needlessly because of human activity, and this simple contribution made me feel like I could help stop an injustice thousands of miles away.

Today, as an avid diver I have been fortunate enough to see the splendor of some of the most pristine and wild places in the world's vast oceans, from the Galápagos Islands to the coral reefs off the coast of Australia.

Through my interest, experiences, and love for the oceans I have developed a special affinity for sharks, the top predators of the seas. Despite their fierce reputation, sharks have become extremely vulnerable to overfishing, and today many species of sharks are nearing permanent extinction. It's hard to comprehend that one of the most ancient creatures on our planet is disappearing, and entirely because of us. The good news is that sharks are still thriving in some of the last untouched parts of our oceans—a sign of ecological health in some areas, and a blueprint for broader conservation efforts everywhere.

Few people ever get to visit the last pristine parts of our oceans—few except for the long-distance fishing fleets that encroach on these wonderful places to harvest their natural resources. I have watched with my own eyes as illegal fishing vessels plunder the richness of marine parks—areas that, despite their protected status, still lack the proper protection and enforcement measures. It may seem like a faceless crime, but industrial-scale fishing is occurring on such a massive scale that it is threatening not only the overall health of the oceans but also our own well-being and long-term survival.

Given all that is at stake, I'm grateful for people like Enric Sala and for all that he is doing to protect the oceans. I am thrilled that my foundation, the Leonardo DiCaprio Foundation, has partnered with his National Geographic Pristine Seas project. Today, less than 2 percent of the ocean is fully protected by law from fishing and pollution. We need to protect so much more. Enric and Pristine Seas have already helped to inspire leaders around the world to protect our oceans with a clear strategy that uses the best of science, exploration, and media. Working together, we plan to rapidly expand the protection of our oceans and all the creatures that reside in them.

Losing the wildest places in our oceans, and on our planet, is not just a crime against nature—it's a crime against every species on Earth. Pristine Seas gives me hope and proves that our oceans can be protected for generations to come. I am proud to be part of these efforts, and we hope you will join us. ◼

Introduction

When I was a little kid growing up on the Mediterranean coast of Spain in the early 1970s, I was spellbound by the underwater world that the French ocean explorer Jacques Cousteau showed us on television. His fearless divers—my heroes—swam among whales and meandered through lush coral reefs full of large groupers and sharks. I dreamed of being a diver on the *Calypso,* Cousteau's famous ship, and exploring remote seas, making countless discoveries along the way.

Yet the seas full of large fish that I saw on TV were a world apart from the Mediterranean of my childhood. Swimming off Catalonia's Costa Brava, all I saw were minnows smaller than my little diving mask. The lush forests were nowhere to be seen; instead, I saw barrens dotted with dark sea urchins. The empty sea was what was natural to me.

Fast-forward 25 years. I was a professor at the Scripps Institution of Oceanography in California, studying the impacts of humans—overfishing, pollution, global warming—on the ocean. Born too late to earn a berth in the *Calypso,* life took me to academia, where I continued my passion to explore the ocean. The initial excitement of using science to understand the world, however, turned into frustration as the places that I loved so much became less and less alive, year after year. I was not only a witness of ocean death, but I was also describing it—using the best available science. I was writing the obituary of the ocean, with more and more precision. That was beyond frustrating, more on the verge of depressing. I felt like a doctor telling a patient the excruciating details of how she is going to die, with no cure available.

What was I supposed to do? The academic system told us to publish scientific papers in scholarly journals, worrying about statistical rigor and experimental design and making peer reviewers happy. But I was not supposed to do anything beyond publishing for an expert audience. A vice chancellor of the University of California, San Diego, distilled that philosophy when he told me and other young assistant professors in 2000 that we should worry only about publishing and teaching the minimum amount of classes, and that "service" (to society) was irrelevant and did not have any weight in academic promotions or salary raises. With my friends Nancy Knowlton and Jeremy Jackson, and other like-minded colleagues, we developed a new graduate program to train students to bridge between marine science, economics, and policy. We wanted new graduates to aim for immediate societal relevance. It was an uphill battle.

I was inside an ivory tower. I was yelling as loud as I could about the threats to the ocean, but very few outside could hear us. And then one day I had one of these moments that materialize only a few times in life, when one would think that the entire

universe conspires to open one's mind to another dimension, to the vision of a dream to be fulfilled.

One morning in October 2000 I walked into my building at Scripps and picked up my mail from the mailboxes on the hall. Among my correspondence there was the current issue of *National Geographic.* I loved receiving the magazine, and I performed a dear ritual every time. I closed the door to my office, sat back on my chair, and put my feet on the desk, overlooking the Pacific Ocean. I gave myself one hour to enjoy the latest stories of explorers from around the world and the most fantastic photographs.

I ripped open the brown envelope that contained the magazine, and opened the issue at random. Large white letters screamed at me from the bottom of the page: "MEGATRANSECT." Above the letters, a double-page, black-and-white photo showed a skinny white man in shorts, wading across a swamp under a thick canopy, followed by a group of overloaded and clearly exhausted Pygmies. I flipped the page and started reading. I could not put it down. That was the first of a three-article series about National Geographic Explorer-in-Residence Mike Fay's walk from Congo to the coast of Gabon, across the wildest tracts of tropical forest in central Africa. He and a group of Pygmies walked for a year and a half, 1,930 kilometers, across jungles with no humans, no houses, no villages, no roads, no fires, nothing. They discovered new populations of gorillas, chimpanzees, and forest elephants, as well as hippos surfing in the sea. Inspired by Mike's discoveries, in September 2002 President Omar Bongo of Gabon announced the creation of 13 national parks, covering 11 percent of Gabon's land and protecting these last jewels of central Africa.

That was my epiphany. Mike gave it to me. Unknowingly, he inspired me to do the same in the ocean. Imagining how that year and a half must have been made me question whether I was living with enough intensity and depth. Therefore, I decided to dedicate the rest of my life to helping bring the ocean closer to its former health and richness. But what was a healthy ocean? And was there any healthy sea left?

Those questions were the beginning of a long quest to find the wildest places in the ocean and help to protect them. In 2005 I led an expedition to the remote northern Line Islands, a very little known archipelago of coral islands and atolls between Hawaii and the Equator. Rising from the deep seafloor are five coral islands, the tops of ancient volcanoes millions of years old. Two of these atolls are uninhabited and free from human impacts. There, we saw for the first time what a healthy ocean is. I can still remember my first dive at the wild Kingman Reef, when a dozen sharks bolted from the deep, surrounding us as soon as we submerged. The bottom of the reef was fully covered by healthy corals of delicate pastel colors. These were like the reefs that Cousteau showed us—on steroids.

In 2007 I left Scripps Institution of Oceanography and, after ten years in American academia, moved back to Spain. The National Research Council (CSIC) offered me a position at a research lab on my beloved Costa Brava, where some of my old-time friends worked. It was an ideal position, for I did not have to teach, supervise students, or participate in academic committees. I could spend all my time working on whatever I chose. For the first time in many years I had the space to think. I learned that serious thinking cannot be achieved here and there, in between planned activities, while one is running on a professional hamster wheel. To be creative, one needs time, a lot of unstructured time, and a lot of reading and asking questions and listening.

One day I asked myself: What would I do if resources were not an issue? What would I do if I weren't afraid? A few weeks later I had the answer. There are still some wild places out there, in the most remote corners of the ocean. I was determined to find them and work with key partners to survey them, show them to the world, and inspire the leaders of the countries to which these pristine areas belong to protect them in large marine reserves—national parks in the sea. I became obsessed with that dream.

In January 2008 I traveled to Washington, D.C., to meet with Terry Garcia and other executives at the National Geographic Society to present my idea. They liked it and agreed to host it at National Geographic. In July 2008 I left Spain's National Research Council and dived into my passion, joining the ranks of National Geographic explorers and starting the Pristine Seas project.

On January 6, 2009, I was at the White House, sitting three rows from President George W. Bush. He was signing a proclamation creating the Pacific Remote Islands Marine National Monument, a behemoth of a marine reserve almost as large as the United Kingdom, which included Kingman Reef and Palmyra Atoll, two of the coral paradises we surveyed back in 2005.

The text of Bush's presidential proclamation highlighted the pristine nature of these coral atolls, and the fact that in these wild places, predators overweigh their prey. We could have never imagined anything like that had we not gone through the impracticality of organizing a research expedition to faraway atolls, had we not ignored warnings from some colleagues that we would never be able to raise funds for exploratory science, had we not been curious about our world.

Pristine places, the last wildernesses of the ocean, are all we have left of the seas of the past. They are the best baselines of what is natural—unlike our personal baselines, which are biased by an ocean degraded by overfishing, pollution, and global warming. These last wild places are virtually unknown except by long-distance fishing fleets, which are starting to target them because they have depleted almost everything else.

We need to save these pristine places before it is too late, before they vanish, unnoticed.

This book tells the story of ten of these places, and how a small group of determined scientists, explorers, photographers, and videographers, together with a few key partners, joined efforts to inspire country leaders to protect them. ▪

STAGHORN CORAL: SIGNS OF RESILIENCE
Palau's pristine reefs, such as these in the Ulong Channel, have come back strong even after being damaged by warming events in the 1990s and a big typhoon in 2012. (page 15)

PRESERVING THE OCEAN'S FUTURE
Enric Sala hovers underwater, examining a World War II shipwreck in Palau. For him, the goal is to save pristine ocean environments before it is too late. (page 17)

The Pristine Reef Food Web

Our understanding of what's natural in marine environments is biased, because we started studying the ocean with modern technologies long after we started exploiting it. Pristine places that have not suffered from direct human impacts give us a chance to see the ocean as it was before humans took over.

Fishing is, historically, the strongest human stressor on ocean eco-systems. The general process of exploitation of a coral reef has been to trim the food web from the top down, first targeting the large predators (for example, sharks, tuna, groupers) and later depleting the lower lev-els of the food web. Pristine places maintain intact food webs, including all the players, from large to small. That includes many thousands of species and hundreds of thousands of predator-prey relationships.

What makes pristine food webs unique is the abundance of large predators. At the top of the food web, they can eat everything below, although they tend to eat fish and marine mammals. Most top predators are threatened or endangered. The Caribbean monk seal is now extinct because of excessive hunting. The Mediterranean monk seal is critically endangered: Only a few hundred remain.

A pristine food web, like the simplified coral reef food web featured opposite, includes top predators plus predators at several levels below. Carnivores eat mostly invertebrates (crabs, worms, and shrimp, for example); planktivores eat plankton, microscopic animals that live in seawater; and herbivores eat plants. Some fish families include species that eat a variety of prey. Damselfish, for example, eat both plankton and plants. Some species spend their time swimming in the water col-umn eating plankton; others eat only algae on the reef. So in the figure opposite, damselfish are colored as both planktivores and herbivores.

The pyramids below compare the biomass, or weight of fish (kilo-grams of fish per square meter), found in a degraded reef and a pristine reef. This shows one of the surprise finds of our Pristine Seas explorations, as we discovered that in pristine reefs, the biomass of the top predators far outweighs the biomass of their prey.

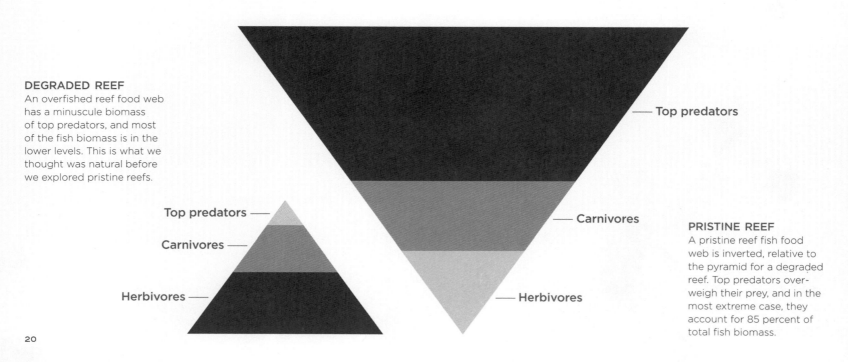

DEGRADED REEF
An overfished reef food web has a minuscule biomass of top predators, and most of the fish biomass is in the lower levels. This is what we thought was natural before we explored pristine reefs.

Top predators ——

Carnivores ——

Herbivores ——

—— Top predators

—— Carnivores

—— Herbivores

PRISTINE REEF
A pristine reef fish food web is inverted, relative to the pyramid for a degraded reef. Top predators over-weigh their prey, and in the most extreme case, they account for 85 percent of total fish biomass.

Top Predators in Pristine Seas

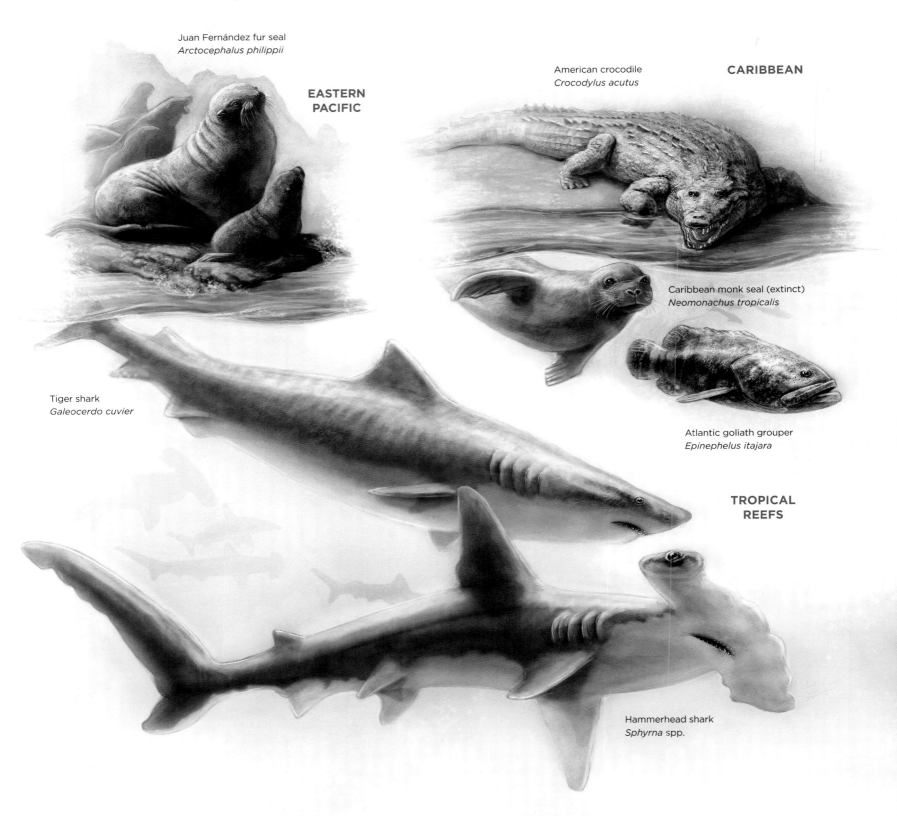

Juan Fernández fur seal
Arctocephalus philippii

**EASTERN
PACIFIC**

American crocodile
Crocodylus acutus

CARIBBEAN

Caribbean monk seal (extinct)
Neomonachus tropicalis

Tiger shark
Galeocerdo cuvier

Atlantic goliath grouper
Epinephelus itajara

**TROPICAL
REEFS**

Hammerhead shark
Sphyrna spp.

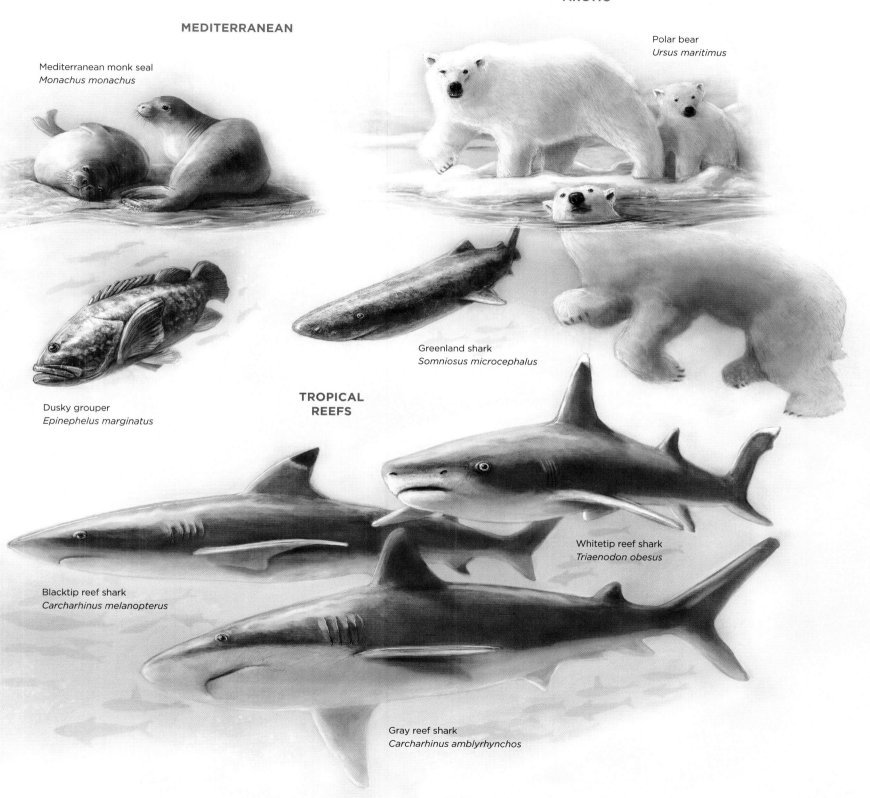

MEDITERRANEAN

Mediterranean monk seal
Monachus monachus

Dusky grouper
Epinephelus marginatus

ARCTIC

Polar bear
Ursus maritimus

Greenland shark
Somniosus microcephalus

TROPICAL REEFS

Blacktip reef shark
Carcharhinus melanopterus

Whitetip reef shark
Triaenodon obesus

Gray reef shark
Carcharhinus amblyrhynchos

Pristine Seas Expeditions

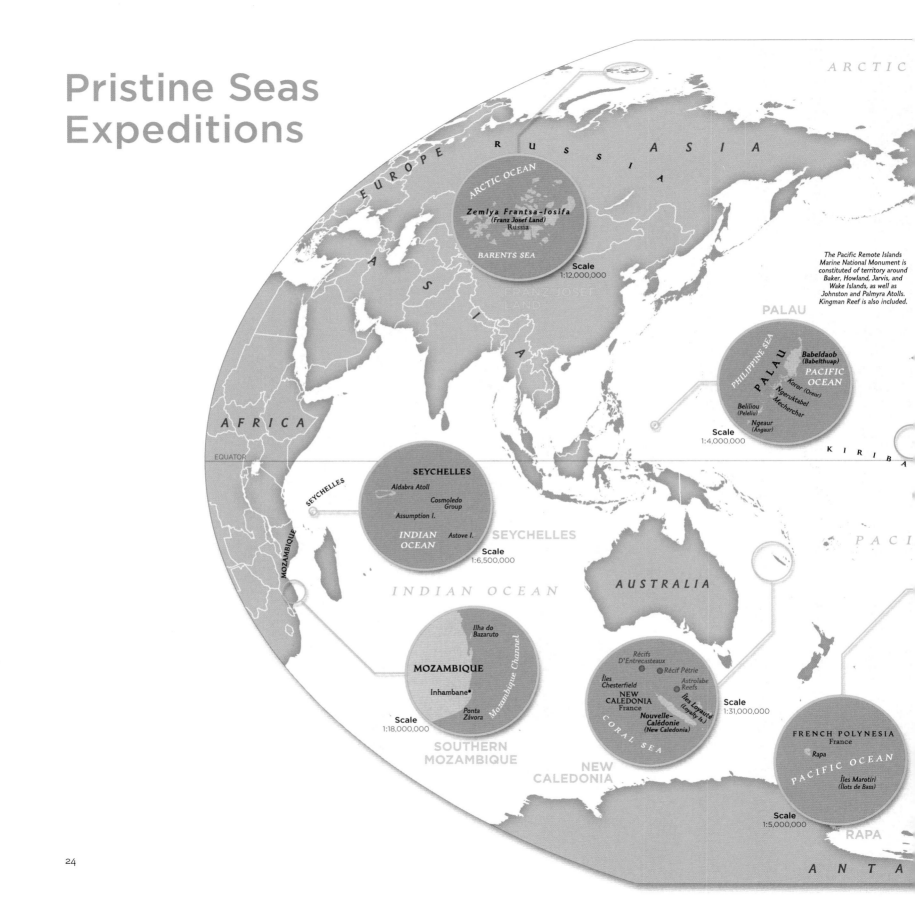

ARCTIC

EUROPE

RUSSIA

ASIA

ARCTIC OCEAN

Zemlya Frantsa-Iosifa
(Franz Josef Land)
Russia

BARENTS SEA

FRANZ JOSEF LAND

Scale
1:12,000,000

A S I A

AFRICA

The Pacific Remote Islands
Marine National Monument is
constituted of territory around
Baker, Howland, Jarvis, and
Wake Islands, as well as
Johnston and Palmyra Atolls.
Kingman Reef is also included.

PALAU

PHILIPPINE SEA

PALAU

Babeldaob
(Babelthuap)

Koror *(Oreor)*
Ngeruktabel
Mecherchar

**PACIFIC
OCEAN**

Beliliou
(Peleliu)

Ngeaur
(Ángaur)

Scale
1:4,000,000

KIRIBA

EQUATOR

SEYCHELLES

SEYCHELLES

Aldabra Atoll

*Cosmoledo
Group*

Assumption I.

Astove I.

**INDIAN
OCEAN**

SEYCHELLES

PACI

MOZAMBIQUE

Scale
1:6,500,000

INDIAN OCEAN

AUSTRALIA

*Ilha do
Bazaruto*

MOZAMBIQUE

Inhambane•

Mozambique Channel

*Récifs
D'Entrecasteaux*

◉ *Récif Pétrie*

*Îles
Chesterfield*

*Astrolabe
Reefs*

**NEW
CALEDONIA**
France

*Îles Loyauté
(Loyalty Is.)*

*Nouvelle-
Calédonie
(New Caledonia)*

CORAL SEA

Scale
1:31,000,000

*Ponta
Závora*

Scale
1:18,000,000

SOUTHERN
MOZAMBIQUE

NEW
CALEDONIA

FRENCH POLYNESIA
France

Rapa

PACIFIC OCEAN

*Îles Marotiri
(Îlots de Bass)*

Scale
1:5,000,000

RAPA

ANTA

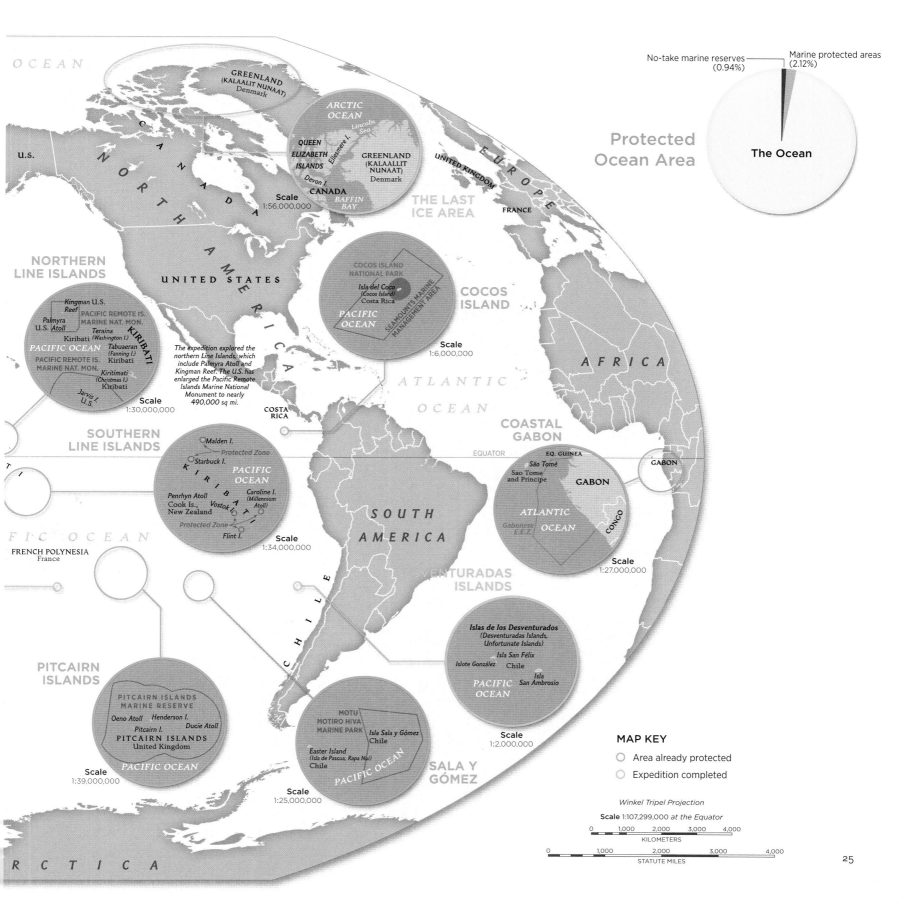

OCEAN

GREENLAND
(KALAALIT NUNAAT)
Denmark

U.S.

CANADA

NORTH AMERICA

UNITED STATES

ARCTIC
OCEAN

Lincoln
Sea

QUEEN
ELIZABETH
ISLANDS

Ellesmere I.

GREENLAND
(KALAALLIT
NUNAAT)
Denmark

Devon I.

CANADA

BAFFIN
BAY

Scale
1:56,000,000

EUROPE

UNITED KINGDOM

FRANCE

THE LAST
ICE AREA

NORTHERN
LINE ISLANDS

Kingman U.S.
Reef

PACIFIC REMOTE IS.
MARINE NAT. MON.

Palmyra
U.S. Atoll

Teraina
(Washington I.)
Kiribati

KIRIBATI

PACIFIC OCEAN

Tabuaeran
(Fanning I.)
Kiribati

PACIFIC REMOTE IS.
MARINE NAT. MON.

Kiritimati
(Christmas I.)
Kiribati

Jarvis I.
U.S.

Scale
1:30,000,000

The expedition explored the
northern Line Islands, which
include Palmyra Atoll and
Kingman Reef. The U.S. has
enlarged the Pacific Remote
Islands Marine National
Monument to nearly
490,000 sq mi.

COSTA
RICA

COCOS ISLAND
NATIONAL PARK

Isla del Coco
(Cocos Island)
Costa Rica

PACIFIC
OCEAN

SEAMOUNTS MARINE
MANAGEMENT AREA

Scale
1:6,000,000

COCOS
ISLAND

ATLANTIC

OCEAN

AFRICA

COASTAL
GABON

SOUTHERN
LINE ISLANDS

Malden I.

Protected Zone

Starbuck I.

KIRIBATI

PACIFIC
OCEAN

Penrhyn Atoll
Cook Is.,
New Zealand

Vostok I.

Caroline I.
(Millennium
Atoll)

Protected Zone

Flint I.

Scale
1:34,000,000

SOUTH
AMERICA

EQUATOR

EQ. GUINEA

São Tomé
Sao Tome
and Principe

GABON

GABON

ATLANTIC
OCEAN

Gabonese
E.E.Z.

CONGO

Scale
1:27,000,000

FIC OCEAN

FRENCH POLYNESIA
France

CHILE

DESVENTURADAS
ISLANDS

Islas de los Desventurados
(Desventuradas Islands,
Unfortunate Islands)

Isla San Félix

Islote González

Chile

Isla
San Ambrosio

PACIFIC
OCEAN

Scale
1:2,000,000

PITCAIRN
ISLANDS

PITCAIRN ISLANDS
MARINE RESERVE

Oeno Atoll

Henderson I.

Pitcairn I.

Ducie Atoll

PITCAIRN ISLANDS
United Kingdom

PACIFIC OCEAN

Scale
1:39,000,000

MOTU
MOTIRO HIVA
MARINE PARK

Isla Sala y Gómez
Chile

Easter Island
(Isla de Pascua, Rapa Nui)
Chile

PACIFIC OCEAN

Scale
1:25,000,000

Scale
1:2,000,000

SALA Y
GÓMEZ

RCTICA

Protected
Ocean Area

No-take marine reserves
(0.94%)

Marine protected areas
(2.12%)

The Ocean

MAP KEY

◯ Area already protected

◯ Expedition completed

Winkel Tripel Projection

Scale 1:107,299,000 at the Equator

0   1,000   2,000   3,000   4,000
KILOMETERS

0   1,000   2,000   3,000   4,000
STATUTE MILES

We were asking the ocean gods to catch a glimpse of a forgotten world, a world that had disappeared from humanity's collective memory.

CENTRAL PACIFIC CORAL REEFS

The Line Islands

The Meaning of Pristine

My office at the Scripps Institution of Oceanography overlooked the Pacific Ocean off La Jolla, California. I used to look out and let my mind wander. There was an infinity of blue out there.

That's what most people know about the sea: It's blue, and it's large. Few put their heads down and see with their own eyes. I had been underwater most of my life, but I knew there was more to see than I had yet discovered. And I still felt consumed by a burning question.

I had become painfully aware that our appetite for seafood was depleting sea life around the world. Human beings are taking fish out of the ocean faster than they can reproduce. The more I learned, the more impassioned I became.

I wanted to bring ocean life closer to what it was before we started to take too many fish out of the ocean. But to bring the ocean back, we need to know how it used to be. And I really did not know. My memory extended 30 years from when I started to swim as a kid in the Mediterranean. I heard the stories from old Mediterranean fishers of how much larger and more abundant fish used to be. But what I saw and what I heard from my elders was just a second in ecological time.

My only way to know what the ocean was like in the past was . . . to travel back in time. My hero National Geographic Explorer-in-Residence Mike Fay virtually traveled back in time by walking through the wildest tracts of forest in central Africa in 1999–2000, and encountering wildlife that had never seen humans before (see Introduction). I just needed to find a pristine ocean swath and dive in.

Palmyra's Promise

One day in 2004 my friend and colleague Jeremy Jackson came into my office, out of breath. Jeremy is a lean giant of a man, with crazy orange hair that he attempts to control in an enormous ponytail and long, knotty fingers that he moves as though playing a piano while he speaks. His blue eyes were wide open, like seeing an apparition. He said: "Enric, you have to go to Palmyra!"

Palmyra is a small coral atoll 1,600 kilometers south of Hawaii. The environmental group the Nature Conservancy raised funds from several private donors and purchased the island from its private owner, to give it to the U.S. Fish and Wildlife Service, in 2000. Jeremy had recently been invited by the Nature Conservancy to visit and dive in the coral reefs around the atoll.

"You have to go to Palmyra and do your study of a healthy coral reef there," said Jeremy. "I jumped in the water and was surrounded by sharks!"

I admired and respected Jeremy so much that I trusted him on the spot, that Palmyra would be the time machine I was looking for, the intact coral reef that had not been destroyed by human beings.

I dropped everything and focused on organizing an expedition to Palmyra Atoll. I hired Stuart Sandin, a recent Ph.D., to help me, and contacted Alan Friedlander, a fish ecologist from the University of Hawaii who knew the reef fish of the region better than anyone else. Alan told us of Kingman, a smaller reef beneath the waves, without permanent emerging land, about 32 kilometers north of Palmyra. That was a place that perhaps could take us further back in time than Palmyra.

To the Northern Line Islands

It was decided. We would go to Kingman and Palmyra, the northernmost of the Line Islands, to describe what coral reefs look like in the absence of people. But I was not satisfied with just looking at fish or corals; I wanted to describe the *entire* reef, from smallest to largest. If we were going so far away, we had better come back with as much information as humanly possible. Therefore, we started to search for 15 of the greatest and brightest experts on everything from viruses to sharks, and chartered for six weeks a rusty former U.S. Navy buoy tender that had been christened on D-Day. I felt like recruiting my own *Oceans 11* gang to rob a fortified vault. Only our booty was not money but knowledge. The donors of my research at the time—individuals and private foundations—were excited about true exploration, and supported the endeavor. And so did National Geographic, with a small grant through its Committee for Research and Exploration.

In August 2005 we sailed to the Line Islands. I didn't know what we were going to find, but we expected it to be extraordinary. Projecting Mike

Fay's jungle adventures onto myself, I hoped to find many sharks and large reef fish that had never seen humans before, to be approached curiously as though we were visitors from outer space. We were asking the ocean gods to catch a glimpse of a forgotten world, a world that had disappeared from humanity's collective memory.

The pristine coral reefs we explored in Palmyra and Kingman cast a spell on me that has lasted since. To paraphrase Jacques Cousteau, I am forever caught in their net of wonder.

Fish That Do Not Swim Away

Vostok is a coral atoll so small that 13 of them would fit in New York's Central Park. When I started to research about Vostok—meaning "east" in Russian—in 2009, there was not even a satellite picture of it. I typed the coordinates of the island on Google Earth, clicked "return," and all that showed up on-screen was a blurry dark triangle in a featureless blue ocean. A few months later, I was sitting on the edge of a rubber boat, donning my diving gear, ready to jump in the water, 30 meters from the shores of Vostok.

It was mind-boggling to try to comprehend that we were on top of what is equivalent to a giant pencil anchored thousands of meters deep on the ocean floor, with only the tip sticking out of the water. Following a ritual I have repeated thousands of times, I adjusted my diving mask, put the diving regulator in my mouth, gave a nod to our underwater cinematographer, Manu San Félix, who was sitting in front of me, and fell backward off board.

A gentle splash on the warm sea was my way of knocking on the door of the underwater world. Nine meters below, thousands of fish swarmed around the coral reef like flocks of multicolored birds. A couple of blacktip reef sharks disappeared among the breaking waves. My heart was racing as if I was about to kiss my first girlfriend. I could not contain my excitement. And then I felt a pull on my ponytail. My diving buddy Alan Friedlander tends to be very funny underwater, and I thought he was teasing me. I turned around and, to my surprise, instead of Alan I saw the face of a half-meter-long twinspot snapper, with fangs projecting out of its mouth, like an underwater vampire. I laughed so hard that the regulator came out of my mouth. The bubbles appeared to scare the snapper, but only for a second. It came right back, hovering a few centimeters from my face. That moment I knew I had realized my dream of being in a place where fish had never seen humans before. I felt how Charles Darwin must have felt the first time he stepped on the Galápagos Islands and realized, to his surprise, that wildlife did not run away from him. The time machine was working perfectly.

Exploring the Southern Line Islands

For the next five weeks we surveyed the unexplored reefs of Vostok, Flint, Malden, Starbuck, and Millennium in the southern Line Islands. We had never seen such healthy coral gardens, so many fish—and sharks—anywhere else. We had found our secret paradise. Of all the places we visited, Millennium is the one that became engraved in our minds.

Millennium Atoll is also known as Caroline Island, but the government of Kiribati renamed it—and changed its time zone—to be the first to enter the new millennium on January 1, 2000. From the air, Millennium is shaped like a banana the size of Manhattan. White foam painted by breaking waves separates the reef and the shallow lagoon from the deep waters around the atoll. A couple dozen islands and islets, each a green oasis ringed by coral rubble, rim the atoll. Within the atoll, the lagoon is what every

computer screen saver dreams of being, the most extraordinary mosaic of blue and turquoise waters. It had a hypnotic quality, for I could not look away. I was mesmerized by the shimmering light, as though being asked by the mermaids to join them beneath the surface.

And so we did. For a week we dived all around the atoll and within its lagoon. We were allowed to enter a precious world, a window into an ocean that has vanished almost everywhere else.

Protecting the Line Islands

We left the Line Islands convinced that our responsibility was not simply to explore and tell the world how magic these last wild places are, but also to help protect them in perpetuity. We published our scientific results, in particular our findings on the inverted biomass pyramid—the fact that in pristine areas, top predators outweigh their prey (see pages 20–21)—and for that we enjoyed a great deal of media attention.

But I also received a few angry emails from people who would prefer these places to remain unknown. My response was—and is—that in the past, remoteness meant de facto protection. Today, unfortunately, there are no remote places anymore. If I could find these pristine places using an old atlas and a little research online, so can pirate fishermen. As a matter of fact, we found shiny hooks in the mouths of gray reef sharks in the southern Line Islands in 2009.

On January 6, 2009, just before the end of his eight-year term, President George W. Bush established the Pacific Remote Islands Marine National Monument, one of the largest no-take marine reserves in history. In 2014 Pristine Seas, partnering with other conservation organizations, inspired President Barack Obama to expand that reserve. On September 25, 2014, he expanded it by more than a million square kilometers, making it the largest protected area network—land and sea—on the planet. ■

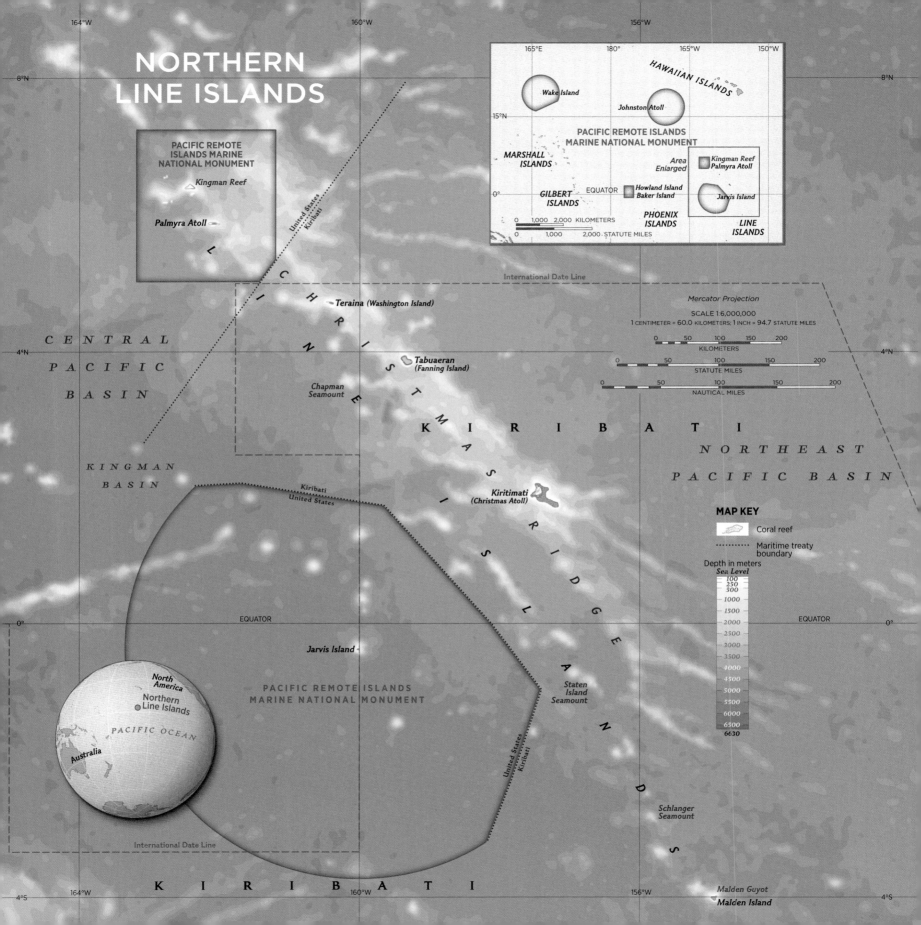

NORTHERN LINE ISLANDS

PACIFIC REMOTE
ISLANDS MARINE
NATIONAL MONUMENT

Kingman Reef

Palmyra Atoll

United States
Kiribati

Inset map (top right)

HAWAIIAN ISLANDS

Wake Island

Johnston Atoll

PACIFIC REMOTE ISLANDS
MARINE NATIONAL MONUMENT

MARSHALL
ISLANDS

Area
Enlarged

Kingman Reef
Palmyra Atoll

EQUATOR

GILBERT
ISLANDS

*Howland Island
Baker Island*

Jarvis Island

PHOENIX
ISLANDS

LINE
ISLANDS

0   1,000   2,000 KILOMETERS
0   1,000   2,000 STATUTE MILES

International Date Line

Teraina (Washington Island)

CENTRAL
PACIFIC
BASIN

*Tabuaeran
(Fanning Island)*

*Chapman
Seamount*

KIRIBATI

Mercator Projection

SCALE 1:6,000,000
1 CENTIMETER = 60.0 KILOMETERS; 1 INCH = 94.7 STATUTE MILES

0   50   100   150   200
KILOMETERS

0   50   100   150   200
STATUTE MILES

0   50   100   150   200
NAUTICAL MILES

KINGMAN
BASIN

Kiribati
United States

*Kiritimati
(Christmas Atoll)*

NORTHEAST
PACIFIC BASIN

MAP KEY

Coral reef

Maritime treaty
boundary

Depth in meters
Sea Level

100
250
500
1000
1500
2000
2500
3000
3500
4000
4500
5000
5500
6000
6500
6630

EQUATOR

Jarvis Island

PACIFIC REMOTE ISLANDS
MARINE NATIONAL MONUMENT

*Staten
Island
Seamount*

North
America

Northern
Line Islands

PACIFIC OCEAN

Australia

United States
Kiribati

*Schlanger
Seamount*

International Date Line

KIRIBATI

Malden Guyot

Malden Island

THE LINE ISLANDS are so named because they spread across "the line"—the Equator. The archipelago extends 2,300 kilometers across the central Pacific Basin and encompasses 11 islands and atolls, which are the only emerging parts of a chain of undersea mountains that rises from a seafloor in excess of 5,000 meters deep. Eight of the islands are part of Kiribati (of which only three are inhabited), and three are territories of the United States. The Line Islands are the first islands to be found west of the American tropics, and they are considered to be stepping-stones for species dispersing across the central Pacific.

GRAY REEF SHARK PATROL
Kingman Reef was the atoll where the Pristine Seas team discovered that, in a pristine coral reef, top predators such as sharks outweigh their prey.

THE SCOURGE OF THE CORALS
A crown-of-thorns sea star crawls over giant clams at Kingman Reef. These sea stars have decimated corals elsewhere, but on pristine reefs their numbers are controlled by their main predator, the giant triton.

TENDING TO THE REEF
Large parrotfish keep the surface
of Kingman Reef clear of seaweed,
which allows corals to replenish
their populations.

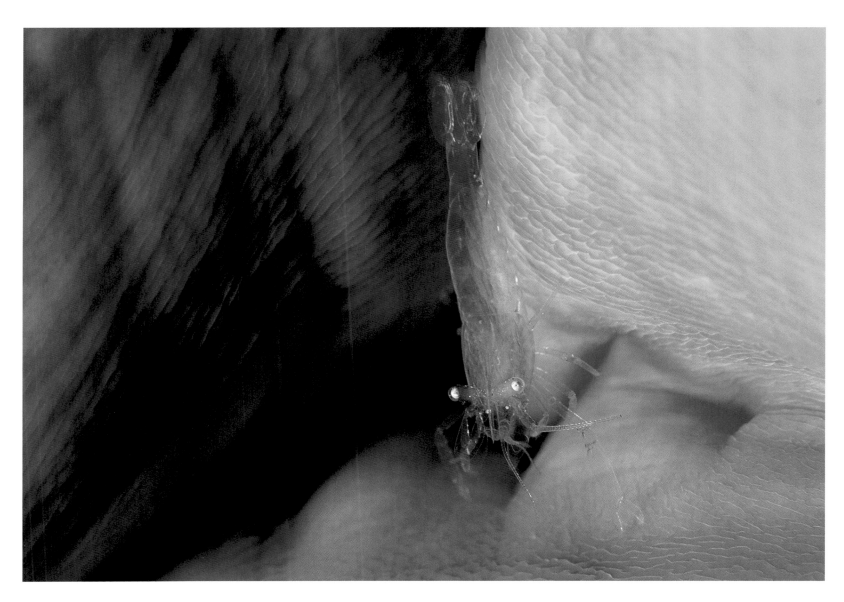

A MINIATURE WORLD OF DIVERSITY
A transparent shrimp the size of a grain of rice nearly
disappears on its anemone host. The microscopic
landscape of the reef is inconspicuous but more
diverse than it might seem.

"The sea, the great unifier,
is man's only hope.
Now, as never before,
the old phrase has a
literal meaning: we are
all in the same boat."

—JACQUES-YVES COUSTEAU

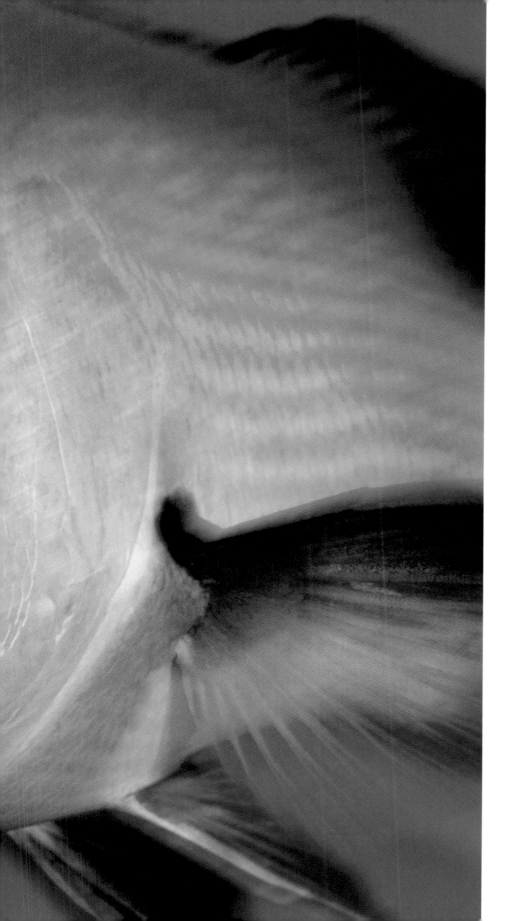

A SNAPPER BARES ITS SHARP TEETH
The twinspot snapper is one of the most abundant predator species on pristine reefs like Kingman. These fish try to eat anything they can get their teeth on—including Enric Sala's ponytail.

THE CONSTANT CRASH OF WAVES
Kingman Reef barely scratches the water's surface, and it is continuously beaten by powerful Pacific swells. Yet corals have been growing and building out the reef for hundreds of thousands of years. *(preceding pages)*

Sharks, Sharks, and More Sharks

When I ask people if they fear anything about the sea, the most common response is "Sharks!" There seems to be a widespread belief that sharks have been evolving in the ocean for 300 million years with the sole purpose of eating every single human that dares to swim in it. Steven Spielberg's movie *Jaws* and the Discovery Channel's Shark Week have not helped to change that perception. But I dive with sharks—with *lots* of sharks, often—and I really miss them when they are not around.

The first time I experienced a truly *sharky* environment was quite comical. We arrived at the wild Kingman Reef on a hot morning in August 2005. Sixteen hundred kilometers south of Hawaii, Kingman looks like a boomerang on satellite photographs. Half of the atoll barely breaks the surface, and the other half is sinking, as ocean volcanoes do after a few million years. We anchored our ship between patch reefs in the lagoon. We were in a sheltered spot, but we could hear the roar of the ocean swell breaking on the outer side of the reef. I was very impatient to dive at the wildest place I had seen in my life, and also a little nervous, precisely because it was so wild. We did the routine check of our diving gear, made sure we were not forgetting any scientific gear or cameras, and jumped into one of our Zodiac inflatable dinghies. We sailed through a passage on the reef, between breakers, and arrived at the most exposed side of the fore reef. My friend Zafer Kizilkaya, the expedition photographer, casually got ready and did a backflip over the rubber side of the Zodiac. Five seconds later, I was about to get into the water myself when Zafer literally jumped, like a dolphin, out of the water and back into the Zodiac.

"There are too many sharks down there, and they are aggressive!" he yelled.

My companions and I looked at each other for what seemed the longest second of my life. We shrugged our shoulders, and I remember saying, "Safety is in numbers. One, two, three!" We all jumped in. As soon as our bubbles cleared, we saw a dozen gray reef sharks swimming around us. They were indeed very curious, but after a few minutes they lost interest in us—weird animals with two tails, throwing bubbles and making a lot of noise—and they went about their business. That was the first of many such shark encounters in the Line Islands. After those dives, I became spoiled for life. Now, when I dive at a place without sharks, I really miss them. ∎

THE ANCIENT SAILOR'S NIGHTMARE
Gray reef sharks are a constant presence in the Pacific Remote Islands Marine National Monument.
But despite their reputation, these predators were never aggressive with the Pristine Seas team.

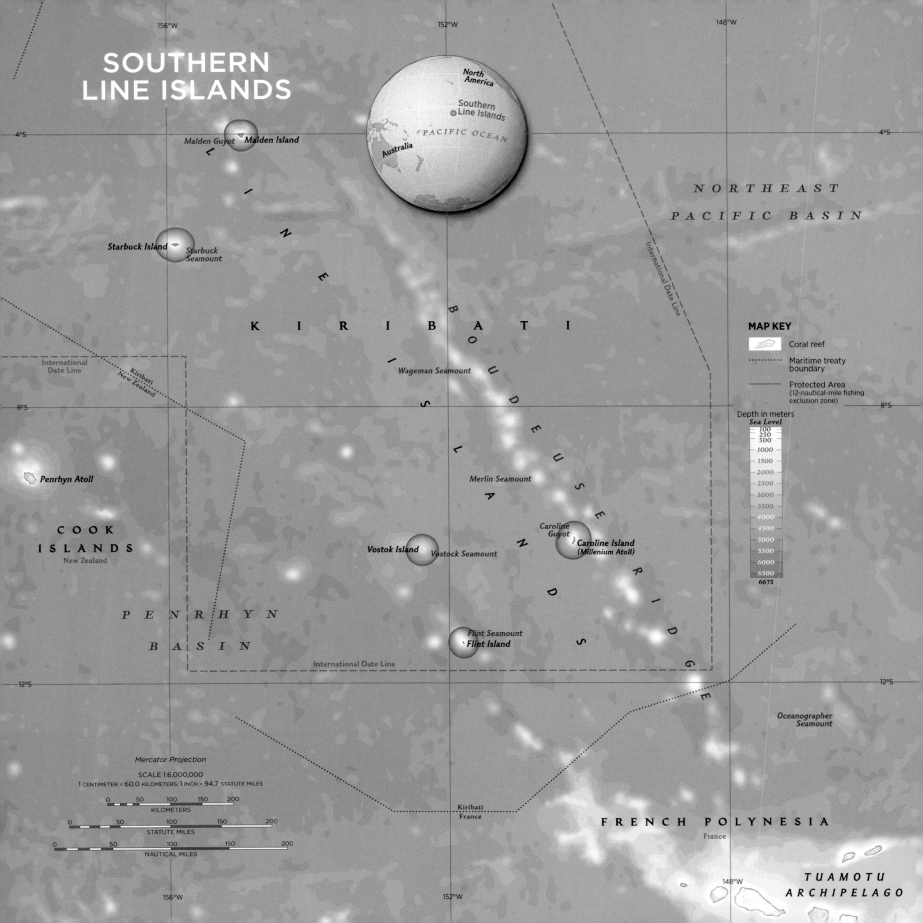

SOUTHERN LINE ISLANDS

North America

Southern Line Islands

PACIFIC OCEAN

Australia

NORTHEAST

PACIFIC BASIN

Malden Guyot **Malden Island**

International Date Line

Starbuck Island *Starbuck Seamount*

K I R I B A T I

Wageman Seamount

International Date Line

Kiribati New Zealand

MAP KEY

Coral reef

Maritime treaty boundary

Protected Area (12-nautical-mile fishing exclusion zone)

Penrhyn Atoll

Merlin Seamount

Depth in meters
Sea Level
- 100
- 250
- 500
- 1000
- 1500
- 2000
- 2500
- 3000
- 3500
- 4000
- 4500
- 5000
- 5500
- 6000
- 6500
6675

C O O K

I S L A N D S

New Zealand

Caroline Guyot

Vostok Island *Vostock Seamount*

Caroline Island
(Millenium Atoll)

P E N R H Y N

Flint Seamount
Flint Island

B A S I N

International Date Line

Oceanographer Seamount

Mercator Projection

SCALE 1:6,000,000

1 CENTIMETER = 60.0 KILOMETERS; 1 INCH = 94.7 STATUTE MILES

0   50   100   150   200
KILOMETERS

0   50   100   150   200
STATUTE MILES

0   50   100   150   200
NAUTICAL MILES

Kiribati France

F R E N C H   P O L Y N E S I A

France

T U A M O T U
A R C H I P E L A G O

THE LINE ISLANDS are divided by the Equator into two dramatically different realms. The northern Line Islands are a perfect gradient of human population, from the uninhabited Kingman Reef to Kiritimati, which had about 5,000 people in 2005, when we conducted our first expedition, allowing us to study the impacts of increasing human population on coral reefs. In contrast, the southern Line Islands are all uninhabited. They were exploited for copra for a short period of time in the late 1800s and early 1900s, but they have not suffered from local fishing. Therefore they provide an invaluable baseline for healthy coral reefs in the central Pacific.

MILLENNIUM ATOLL: A PERFECT CORAL REEF
Healthy corals covering every inch of the seafloor and numerous gray reef sharks patrolling are an extremely rare sight elsewhere. A thousand years ago, this is what reefs may have looked like.

A SAFE HAVEN FOR SEA TURTLES
A green turtle swims safe from poachers in the waters around Millennium Atoll. Turtles nest unbothered on the remote beaches of the southern Line Islands, whereas in most other places they are killed for meat and their eggs are collected. *(following pages)*

"If all the beasts were gone, men would die from a great loneliness of spirit, for whatever happens to the beasts also happens to the man. All things are connected."

—CHIEF SEATTLE OF THE SUQUAMISH INDIANS

CURIOUS TEENAGE SHARKS
In pristine areas sharks tend to inspect the odd human visitor carefully. These young blacktip reef sharks surrounded the Pristine Seas team as we waded across a reef flat in the lagoon of Millennium Atoll.

PLAYING CAT AND MOUSE—OR SHARK AND SMALL FISH
A school of convict tangs hugs a reef covered by pink encrusting algae at Millennium Atoll, shying away from their predator—a blacktip reef shark.

The Giant Clams

It took us an hour to negotiate the shallow reefs of the lagoon of Millennium Atoll. We put our masks on, adjusted our snorkels, and slid into the crystal clear water with care, as though trying not to disturb the peace of the underwater world. Two meters below us was a bed of giant clams, each the size of a great big melon. The giant clams were crowded—up to 20 per square meter—and each pair of curved shells harbored a delicate animal with large lips of impossible colors that spilled over the edge of the shell. I drifted with the weak current above the pulsating kaleidoscope of electric blues and greens. It was like a hallucinogenic trip—without the drugs. As I swam above them, I cast a shadow. They closed their shells abruptly, a natural reaction to predators. After a few seconds, they opened their shells again, timidly.

We were interested in the giant clams because the day before microbiologist Forest Rohwer and his team had analyzed the lagoon and found the cleanest water they had ever measured in a coral reef. The concentration of microbes here was ten times lower than at any inhabited reef in the Pacific. Thinking that giant clams had something to do with it, Forest had come back to borrow a couple to conduct an experiment.

Aboard our expedition ship, the *Hanse Explorer,* Forest filled three tanks with water from the lagoon. In one tank he placed a living giant clam; in another, an empty giant clamshell. The third tank he left empty as an experimental control.

Forty-eight hours later, he took water samples from each tank. The result was striking. In the tank with the living clam, microbes were as rare as at the start of the experiment; in contrast, the microbe populations had surged tenfold in the other two tanks. Counting the microbes under the microscope was not even necessary, for the tanks without living clams were cloudy.

Forest then tried something else. He added known concentrations of *Vibrio*—a type of bacteria that causes dysentery and cholera—to each tank. Two days later, the living giant clam had filtered the water and made *Vibrio* undetectable, whereas in the two other tanks the abundance of *Vibrio* had skyrocketed. This simple yet ingenious experiment clearly showed that giant clams are not only beautiful animals but also natural filters that keep the reefs—and us—healthy. ■

A PRISTINE GIANT CLAM BED AT MILLENNIUM ATOLL
Giant clams carpet the reefs in the Millennium lagoon, a rare sight these days.
Most clam populations in the Pacific have been decimated by human exploitation.

This was an evolutionary dream, the product of millions of years of biological trial and error . . . Far from man, nature is more spectacular.

SUBTROPICAL REEFS

Pitcairn Islands and Sala y Gómez

The Bounty of the Sea

"Grab this light and follow me," said Edmundo Edwards one night in February 2011. He closed the door of his car with a gentle thump, and started walking. Edmundo is a modern Indiana Jones, including the brown uniform. His beard and glasses make him look like a university professor, and his gentle and humble demeanor masks his encyclopedic knowledge of the human history of Polynesia. We were stepping on sacred grounds of the ancient Rapa Nui—as the inhabitants of Easter Island call their island and themselves. "Here, point the light from that side of the rock. Lower," instructed Edmundo. As I lowered my light and projected it sideways onto a flat brown rock, a figure emerged, a large tuna. A few feet away, another rock carving revealed a sea turtle. Those messages from the past told of a strong connection between ancient Polynesian peoples and the sea.

Rapa Nui is one of the most remote inhabited islands on the planet, famous for its giant statues (moai) chiseled off volcanic rock, its complex pre-European society, and the environmental and societal collapse that followed the overexploitation of the island's natural resources centuries ago. When the first European visitors arrived in 1722, there were no trees taller than three meters. Scholars argue about the ultimate cause of deforestation, but in any case, there is agreement that human activities turned Rapa Nui into a microcosm of the world's environmental problems.

Rapa Nui's Untouched Neighbor

In 2008, when I was looking for pristine places to explore, I could not find much information about the state of the underwater ecosystems of Rapa Nui. The island did not show up in my global analysis as "pristine," and I assumed that the environmental degradation on land had been mirrored in the sea. But 200 nautical miles east of Rapa Nui, a bright spot stood out: Sala y Gómez. When I was a kid, one of my favorite books was an oversize gray hardback, with its large "World Atlas" title written in gold. I spent many an hour browsing those giant pages, fascinated by all those exotic locales that I dreamed of visiting one day. I remember especially well one particular island, because it bore my name: Sala. When the island showed up as potentially pristine in my analysis, it took me one nanosecond to decide we were going to go.

In 2010 and 2011 we conducted two expeditions to Rapa Nui and Sala y Gómez. We discovered that fishing has depleted the Rapa Nui waters, whereas Sala y Gómez remains as pristine as when the old Rapa Nui carved magnificent sea animals on volcanic rock (see pages 76–77).

A Mariner's Dream

I was trapped on an island, surrounded by thousands of kilometers of unforgiving sea. I had no ship, and no ships were coming. I was isolated from the world, from everything I ever loved and longed for, in a barless prison, on top of a dark rock rimmed by precipitous cliffs. The waves beat the island incessantly, eroding it little by little, and filling the air with an atrocious sound. I could not escape; I lost all hope I'd ever come back home. My heart drummed, panic overtook my mind, and then . . . I woke up, sopping wet sitting in bed, surrounded by silence, in the middle of a serene night. The moon had painted the sea with a delicate silver stroke, and a light wind caressed the palm trees outside of my window. I was on an island, but that was no prison. I was there by my own will.

My nightmare transported me to 1790, when Fletcher Christian and a handful of British mutineers arrived on Pitcairn Island in the South Pacific, fleeing from His Majesty's justice, after having taken possession of the infamous H.M.S. *Bounty* (see pages 90–91). We are told that Christian spent a long time thinking about what they did, looking at the sea from a cave on the cliffs. Maybe I caught a glimpse of his memory during my own first night on the island.

The human history of the most famous mutineers is the most pervasive memory of Pitcairn. But we were there for something bigger, something with a longer history. We wanted to tell a different story of this part of the world. Our mission was to explore the underwater world of Pitcairn and three uninhabited islands nearby, a world that almost no one has seen—and to explore with the local population a future for their seas.

The four islands—Pitcairn, Ducie, Henderson, and Oeno—form one of the most remote archipelagoes on the planet. Pitcairn is inhabited by just over 50 people, most of them descendants of the mutineers of the *Bounty*. The other three islands are uninhabited. Even if you desire to travel to these islands, you will have a hard time. There are no airports, and only one ship travels to Pitcairn on a regular schedule—four times a year. In fact, the Apollo astronauts made it to the moon faster than it took us to go from Washington, D.C., to Pitcairn. Traveling to Pitcairn is traveling back to a time where things take longer, where everything comes into deeper perspective. Out there, one realizes what is truly important.

From Tahiti we took a weekly flight to Mangareva, an atoll in the Gambier Islands, the southeasternmost archipelago in French Polynesia. The *Claymore II,* a ship based in New Zealand, was waiting for us there. It then took a night, a day, and another night to reach Pitcairn. During the crossing our ship rolled ceaselessly, climbing up and down swells large as a two-story house. These were long swells, coming all the way from Antarctica, where giant storms unleash the power of the ocean. These swells can be felt as far north as Hawaii, to the delight of surfers. In our case, they made us feel miserable. Our stomachs and heads did not want any part of it, and we chose to hunker down and hibernate.

Flying Underwater

The morning of the second day we saw Pitcairn rise out of the dawn. The sky was the color of lead, and the island was dark and imposing, like an impenetrable fortress. As we got closer, the black became greens and browns and reds. Large chunks of the island had fallen into the sea. The local Pitcairners would later tell us that a month earlier they suffered the heaviest rains in many decades. In just one day they received as much rainwater as they did in the entire previous year. The islands could not take it, and some of that land and its precious soil cascaded into the sea. Because of that, Pitcairn was ringed by a halo of brown water. Visibility was less than one meter, and scientific work and filming seemed impossible under such circumstances. Waves kept pounding the island's shores, and the shallow waters became dangerous for diving.

Trying to make the best of the situation, we decided to travel to Ducie Atoll. It took us a day and a half to reach Ducie. The weather improved, and for a few days we had sunny blue skies, which contrasted with a dark blue sea. Ducie is the top of an ancient volcano that emerged from the seafloor eight million years ago, at the time when the human lineage and chimpanzees separated from a common ancestor. Yet modern humans would not reach Ducie until 1606. Even

with a ship and on a calm day, it is not easy to see Ducie. From a few kilometers away, Ducie is only a slim thickening of the horizon. Its highest elevation is less than five meters. No wonder it took so long to find it.

The water at Ducie reached another level of hyperbole. I had never seen such pure blue, clear water. We could still see each other underwater at a distance of 75 meters!

Diving at Ducie, we entered paradise. As far as we could see there was a landscape with smooth rolling hills, a pristine coral reef, with pale blue corals looking like giant roses covering the entire bottom. Man could never dream anything like that. This was an evolutionary dream, the product of millions of years of biological trial and error. This purity was also the consequence of isolation. Far from man, nature is more spectacular.

When we recovered from the hallucination-inducing pale blue reef, we realized we were surrounded by thousands of *nanwe*—the Pitcairn name

for chub. The nanwe eat algae, and they are supposed to swim near the bottom. But the nanwe at Ducie were clearly unaware of today's human assessment and were darting to the surface in unison and then swimming back to the bottom, like roller-coaster riders!

Among the nanwe we saw our first sharks, gray reef sharks. As they swam toward us, holes opened on the nanwe clouds, not too frantically under the presence of the ultimate predator, but with great stealth, as though they knew the sharks were more interested in checking us out than in eating them.

Sharks Draw Near

The sharks were very curious, and they came on almost every dive to check us out, on occasions so closely that they bumped the domes of our cameras, yet we never felt threatened by them. Sharks are the ultimate predators, and they know that the secret to a long life is not to be reckless. While young sharks act like teenagers, cocky and goofy, large sharks are cautious and they approach divers very slowly. They circle you forever, drawing an infinite spiral from the outside in, which never seems to reach the center. And brisk movements scare them, so the secret to experiencing the beauty of sharks is to keep calm.

After Ducie we sailed to Henderson, the last of its kind, the only raised atoll that harbors a pristine forest, including four species of birds that live nowhere else on the planet.

Sharks at Henderson were more curious than those at Ducie. They seemed to love to bump onto our camera domes, and they scratched mine beyond repair. Yet we stayed calm and surrendered to the sense of awe and wonder that comes with observing an intact marine ecosystem. My heart was beating slowly and rhythmically, and I had goose bumps underneath my wet suit. Other than love, this is what makes me happiest and most filled with life. ∎

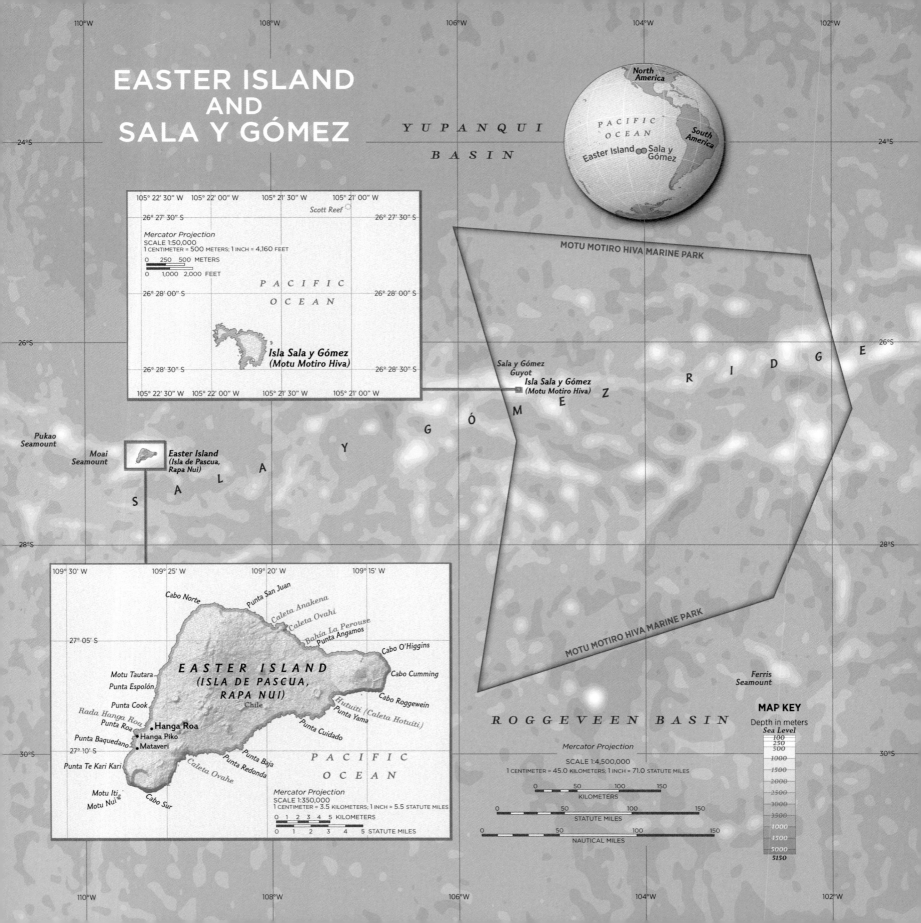

EASTER ISLAND
AND
SALA Y GÓMEZ

YUPANQUI

BASIN

PACIFIC
OCEAN

North
America

South
America

Easter Island ⊙⊙ Sala y
Gómez

MOTU MOTIRO HIVA MARINE PARK

105° 22' 30" W 105° 22' 00" W 105° 21' 30" W 105° 21' 00" W

26° 27' 30" S   26° 27' 30" S

Scott Reef

Mercator Projection
SCALE 1:50,000
1 CENTIMETER = 500 METERS; 1 INCH = 4,160 FEET

0   250   500 METERS

0   1,000   2,000 FEET

PACIFIC

26° 28' 00" S   26° 28' 00" S

OCEAN

26°S

Isla Sala y Gómez
(Motu Motiro Hiva)

26° 28' 30" S   26° 28' 30" S

105° 22' 30" W 105° 22' 00" W 105° 21' 30" W 105° 21' 00" W

Sala y Gómez
Guyot

Isla Sala y Gómez
(Motu Motiro Hiva)

R I D G E

26°S

Pukao
Seamount

Moai
Seamount

Easter Island
(Isla de Pascua,
Rapa Nui)

S A L A   Y   G Ó   M   E   Z

28°S

MOTU MOTIRO HIVA MARINE PARK

28°S

109° 30' W   109° 25' W   109° 20' W   109° 15' W

Cabo Norte

Punta San Juan

Caleta Anakena

Caleta Ovahi

Bahía La Perouse

27° 05' S

Punta Angamos

Cabo O'Higgins

EASTER ISLAND
(ISLA DE PASCUA,
RAPA NUI)

Cabo Cumming

Motu Tautara
Punta Espolón

Cabo Roggewein

Punta Cook

Chile

Hutuiti (Caleta Hotuiti)
Punta Yama

Rada Hanga Roa
Punta Roa

Hanga Roa

Punta Cuidado

Punta Baquedano
Hanga Piko
Mataveri

27° 10' S

Punta Baja

PACIFIC

Ferris
Seamount

Punta Te Kari Kari

Caleta Ovahe
Punta Redonda

OCEAN

MAP KEY

30°S

Motu Iti
Motu Nui

Cabo Sur

Mercator Projection

R O G G E V E E N   B A S I N

SCALE 1:350,000
1 CENTIMETER = 3.5 KILOMETERS; 1 INCH = 5.5 STATUTE MILES

0 1 2 3 4 5 KILOMETERS

0 1 2 3 4 5 STATUTE MILES

Mercator Projection

SCALE 1:4,500,000
1 CENTIMETER = 45.0 KILOMETERS; 1 INCH = 71.0 STATUTE MILES

0   50   100   150
KILOMETERS

0   50   100   150
STATUTE MILES

0   50   100   150
NAUTICAL MILES

Depth in meters
Sea Level

100
250
500
1000
1500
2000
2500
3000
3500
4000
4500
5000

5150

EASTER ISLAND AND SALA Y GÓMEZ,
the southeasternmost lands in Polynesia, are found
410 kilometers apart in the Pacific. Easter Island
(Rapa Nui) is one of the most isolated inhabited islands in the world and is famous for its large
stone statues, its complex pre-European society,
and the cultural and environmental collapse it evidences, likely due to the overexploitation of its natural resources centuries ago. Sala y Gómez (Motu
Motiro Hiva), to the east, is a very small uninhabited
island, 700 meters long with a surface area of only
0.15 square kilometer. An undersea mountain ridge
connects Sala y Gómez and Easter Island.

THE PRISTINE WATERS OF SALA Y GÓMEZ
Galápagos sharks and black trevallies roam
the clear waters in search of small fish prey.
Easter Island, only 200 nautical miles away,
has been fished for centuries and such a
scene is no longer possible there.

CORAL AS OLD AS METHUSELAH
This giant coral colony was alive before the
first European explorers arrived at Easter
Island. Located at the far edge of a region
with good environmental conditions for coral
growth, these corals are extremely healthy
and abundant. *(following pages)*

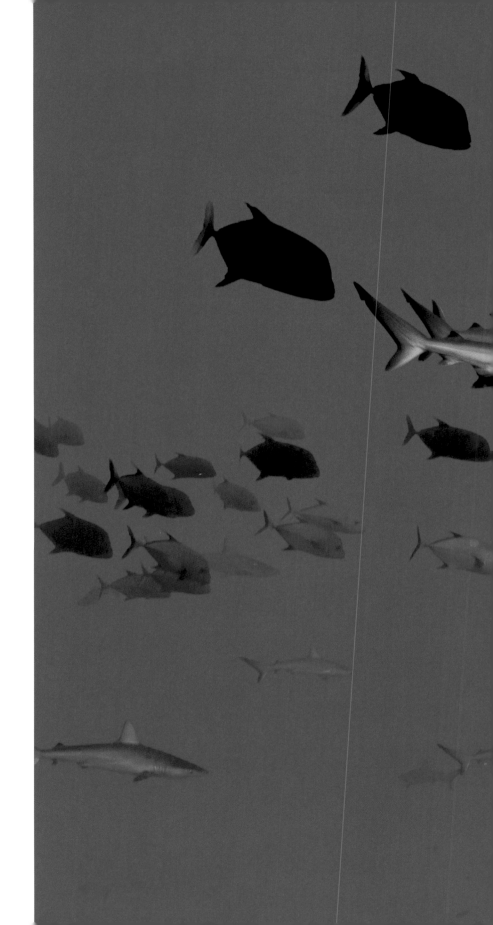

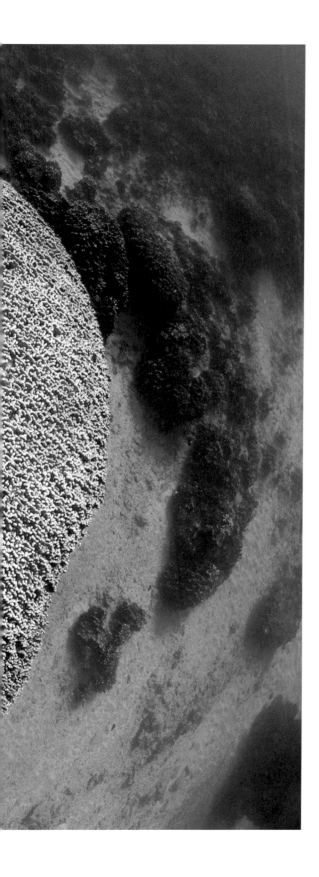

"There is pleasure in the pathless woods, there is rapture in the lonely shore, there is society where none intrudes, by the deep sea, and music in its roar; I love not Man the less, but Nature more."

—GEORGE GORDON, LORD BYRON

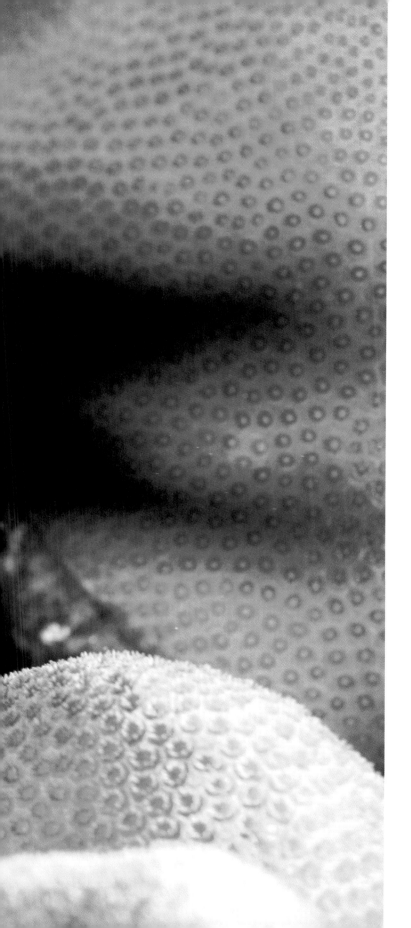

FEARSOME BUT SHY
Morays have a bad reputation, but they really are quite shy, like this stout eel on Easter Island. They typically hide in crevices or search for small prey in between coral branches.

BELOW THE BREAKING WAVES AT SALA Y GÓMEZ
Chubs are a preferred food fish on Easter Island,
where locals eat their intestines raw on the boat,
right after pulling them from the water. These swim
below the breakers of Sala y Gómez, as if flying
through white clouds.

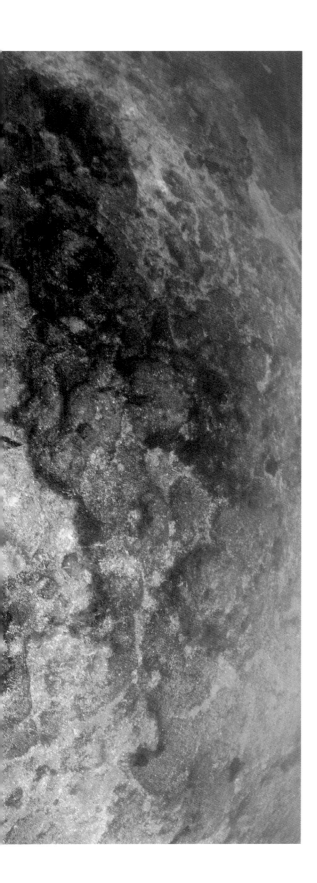

THE WORLD ON THE BACK OF A TURTLE
Green sea turtles surface daily at the little port of Hanga Roa on Easter Island, where local children try to ride them.

A Window to the Past

The morning was chilly, and the sky was made of delicate pinks and blues. The wind was strong enough to raise some whitecaps on the distance, but we were 800 meters off the leeward side of a small strip of rock in the middle of the Pacific Ocean. Our ship rocked rhythmically, like the horses on a carousel, lifted by swells that came all the way from Antarctica, building energy for 2,000 nautical miles, without anything in between. When those swells find an obstacle, their energy explodes in violent waves. That was my first sight of Sala y Gómez: a 700-meter-long, thin black rock in the middle of nowhere, haloed by the white foam of that exploding energy. It was an unforgiving, unwelcoming place, yet we were going to take our chances and dive into that chaotic sea. From the ship *Comandante Toro,* which the Chilean Navy graciously provided for our expedition, a navy crew took us on military speedboats as close as they could to the breakers. One, two, three, and my diving companions and I did backflips and dived down fast, to avoid the washing machine near the surface. We had been diving near Easter Island (Rapa Nui) the week before and, to our surprise, found very healthy reefs with enormous coral heads and canyons—but they looked like empty cities, without the fish one would expect to see. In more than a hundred hours of diving at Rapa Nui, we did not see a single lobster either. Our hope was that Sala y Gómez—uninhabited and unfished—would be closer to what Rapa Nui was before intense fishing. Ten meters deep we were still pulled up and down by the swell, but at least we could control our movements. I looked up. The breaking waves were cotton white cumulonimbus clouds beneath which a large yellowtail jack, a couple of sharks, and a school of chubs seemed to be flying. Twenty meters below, our Rapa Nui diving veteran, Michel García, took us into a cave. We turned our lights on and dived in. I pointed my light to one dark corner. A few large Easter Island spiny lobsters moved nervously. Next to them was a slipper lobster, the most primitive looking of all the lobsters. My friend and colleague Kike Ballesteros measured them. Back on the ship we compared Kike's measurements with those in the scientific literature and found that the lobsters we saw were a third larger than the maximum size reported. I could think of only one reason: Previous studies were conducted in exploited waters, where fishing drives the abundance and size of organisms down. Ours were probably the first measures at the intact Sala y Gómez. We were resetting the scientific baseline. We found our window to the past, a proxy for what Rapa Nui must have been. ■

A TUNA CARVED INTO STONE
Ancient art evidences the Rapa Nui people's ancestral connection to the ocean.
Petroglyphs on Easter Island portray tuna, turtles, and the birdman Tangata Manu.

PITCAIRN ISLANDS

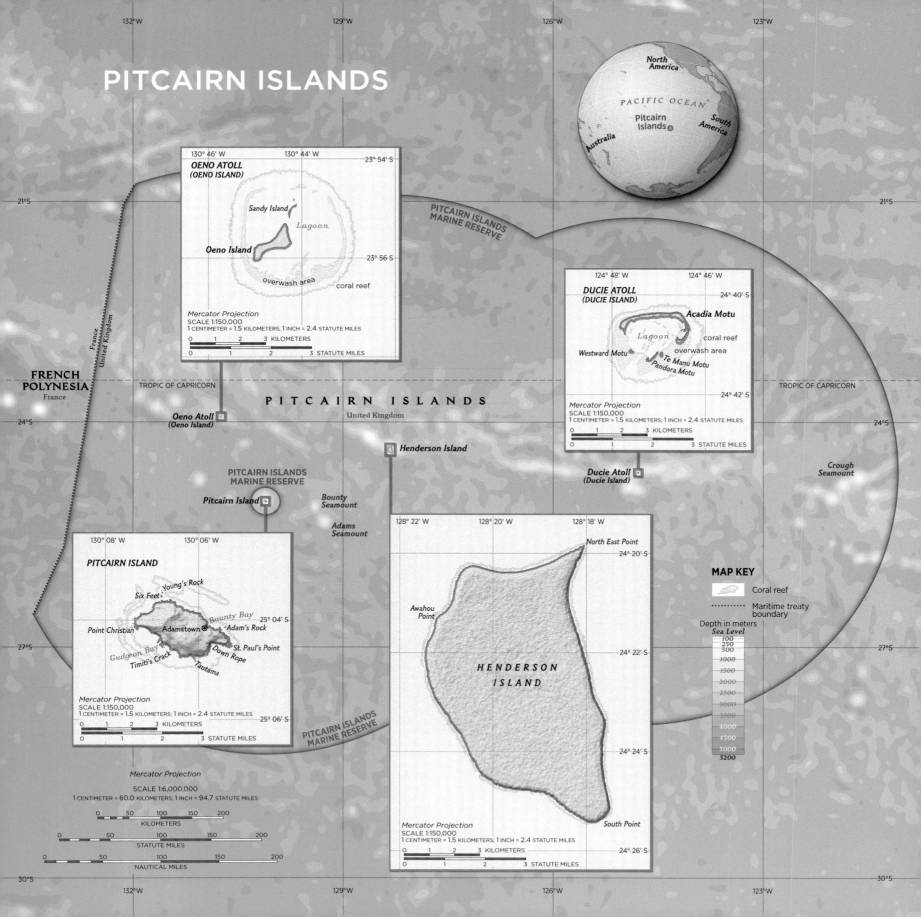

North America

PACIFIC OCEAN

Pitcairn Islands

Australia

South America

OENO ATOLL
(OENO ISLAND)

130° 46' W   130° 44' W   23° 54' S

Sandy Island

Lagoon

Oeno Island

23° 56' S

overwash area   coral reef

Mercator Projection
SCALE 1:150,000
1 CENTIMETER = 1.5 KILOMETERS; 1 INCH = 2.4 STATUTE MILES

0   1   2   3 KILOMETERS

0   1   2   3 STATUTE MILES

PITCAIRN ISLANDS
MARINE RESERVE

DUCIE ATOLL
(DUCIE ISLAND)

124° 48' W   124° 46' W   24° 40' S

Acadia Motu

Lagoon   coral reef

overwash area

Westward Motu

Te Manu Motu
Pandora Motu

24° 42' S

Mercator Projection
SCALE 1:150,000
1 CENTIMETER = 1.5 KILOMETERS; 1 INCH = 2.4 STATUTE MILES

0   1   2   3 KILOMETERS

0   1   2   3 STATUTE MILES

21°S

**FRENCH
POLYNESIA**
France

France
United Kingdom

TROPIC OF CAPRICORN

P I T C A I R N   I S L A N D S

United Kingdom

TROPIC OF CAPRICORN

24°S

Oeno Atoll
(Oeno Island)

Henderson Island

Ducie Atoll
(Ducie Island)

*Crough
Seamount*

PITCAIRN ISLANDS
MARINE RESERVE

Pitcairn Island

*Bounty
Seamount*

*Adams
Seamount*

PITCAIRN ISLAND

130° 08' W   130° 06' W

Young's Rock

Six Feet

Bounty Bay

Point Christian

Adamstown ◉   *Adam's Rock*

25° 04' S

Gudgeon Bay   *St. Paul's Point*

Timiti's Crack   *Down Rope*

Tautama

25° 06' S

Mercator Projection
SCALE 1:150,000
1 CENTIMETER = 1.5 KILOMETERS; 1 INCH = 2.4 STATUTE MILES

0   1   2   3 KILOMETERS

0   1   2   3 STATUTE MILES

128° 22' W   128° 20' W   128° 18' W

North East Point

24° 20' S

Awahou
Point

HENDERSON
ISLAND

24° 22' S

24° 24' S

South Point

24° 26' S

Mercator Projection
SCALE 1:150,000
1 CENTIMETER = 1.5 KILOMETERS; 1 INCH = 2.4 STATUTE MILES

0   1   2   3 KILOMETERS

0   1   2   3 STATUTE MILES

PITCAIRN ISLANDS
MARINE RESERVE

Mercator Projection

SCALE 1:6,000,000
1 CENTIMETER = 60.0 KILOMETERS; 1 INCH = 94.7 STATUTE MILES

0   50   100   150   200
KILOMETERS

0   50   100   150   200
STATUTE MILES

0   50   100   150   200
NAUTICAL MILES

MAP KEY

Coral reef

Maritime treaty
boundary

Depth in meters
Sea Level
100
250
300
1000
1500
2000
2500
3000
3500
4000
4500
5000
5200

27°S

30°S

132°W   129°W   126°W   123°W

PITCAIRN ISLAND is perhaps best known as the home of the descendants of the infamous H.M.S. *Bounty* mutineers. The Pitcairn Island group consists of four of the most remote islands in the world (Pitcairn, Henderson, Oeno, and Ducie), which together form the last remaining British overseas territory in the Pacific. These islands represent the top of ancient volcanoes, located about 2,000 kilometers southeast of Tahiti and 1,900 kilometers west of Easter Island. Together they encompass only 43 square kilometers of emergent land, but the 200-nautical-mile exclusive economic zone that surrounds them covers 836,108 square kilometers, more than three times the size of the United Kingdom.

BLACK TREVALLY SILHOUETTES IN THE WATER
Protected from fishing at Henderson, one of the Pitcairn Islands, these fast predators aggregate in schools to terrorize small fish as they move around the island's coast.

THE GOLD OF THE REEF
The lemonpeel angelfish lives among coral branches and
eats small invertebrates such as crabs. Its distinctive bright
yellow color and blue eye markings make it the target of
live reef fish fisheries across the Pacific.

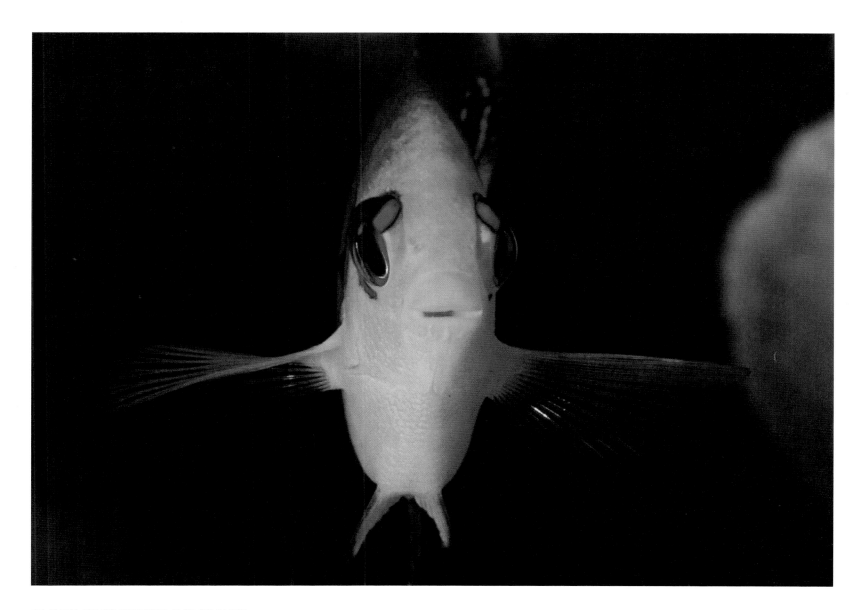

AN ANGEL FOUND NOWHERE ELSE ON EARTH
The Hotumatua is a species of angelfish found nowhere else but in
the Pitcairns, Easter Island, and French Polynesia. These shy fish dart
in and out of coral heads, nervous in the presence of divers and
fearful of being eaten by the many predators on the reef.

A GARDEN OF GIANT BLUE ROSES
The coral *Montipora aequituberculata* forms healthy, immaculate reefs that extend as far as the eye can see in the crystal clear waters off Ducie Atoll.

UNLIKELY COUPLE
At Oeno Atoll, a slender cornetfish (top) shadows a yellow-edged lyretail as if linked by a powerful magnetic field. Smaller carnivores sometimes follow larger ones, hoping to catch any prey that escapes. *(following pages)*

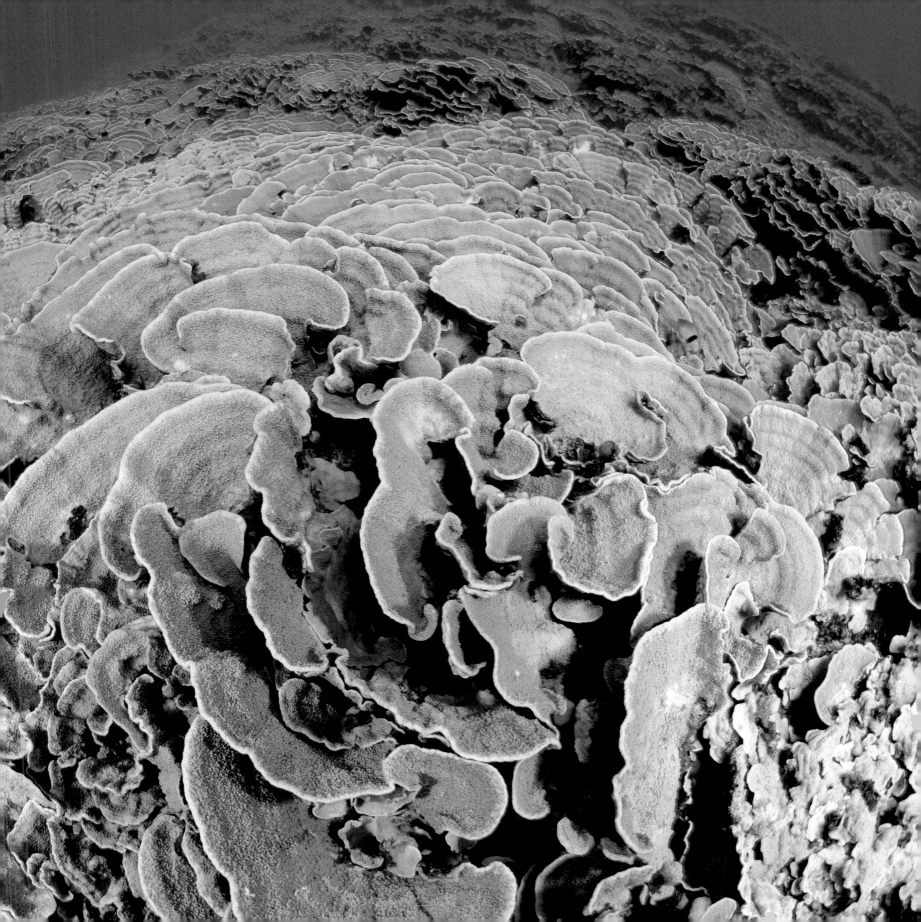

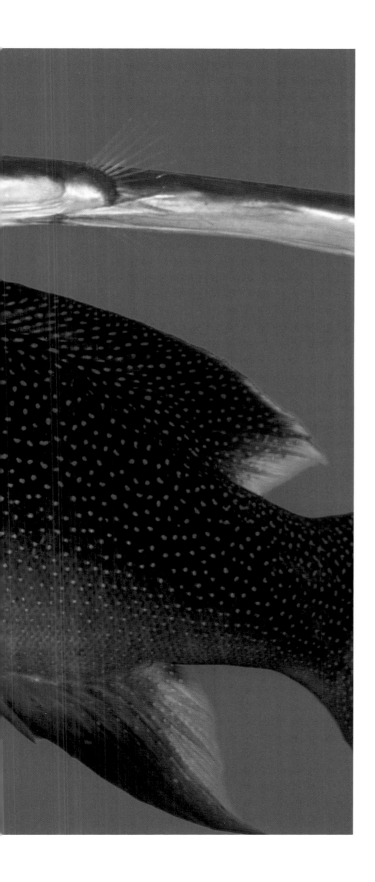

"It's more important to me to revive something that is disappearing than to build something that can be duplicated."

—LARS-ERIC LINDBLAD

THE FLEXIBLE PREDATOR
Whitetip reef sharks, like this one at Ducie Atoll, have a compressed and highly flexible body that allows them to enter small holes and crevices on the reef in search of prey.

The Mutiny and Its Legacy on Pitcairn

Depending on your age, you will associate the story of the mutiny on the *Bounty* with actors Clark Gable, Marlon Brando, or Mel Gibson. They each starred as Fletcher Christian in one of three versions of a popular movie, *Mutiny on the Bounty,* based on a historic event that occurred near Pitcairn Island, this place so far from civilization. Visiting the island, I thought back on the story. It took on even more striking meaning once I saw how isolated the circumstances must have been.

In 1790 the real Christian and a handful of fellow British mutineers arrived here. Under the command of Captain William Bligh, the *Bounty* had been ordered to sail to Tahiti to fetch young breadfruit plants and bring them to the West Indies, where they would be established as a crop to help feed the slave population there. But on April 28, 1789, growing tired of Captain Bligh's harsh treatment of them—and also, perhaps, reminiscing about their five months in idyllic Tahiti in the company of beautiful women—Christian and his companions took over the ship. Christian set Bligh and some of his officers afloat in a small boat, and then he and his fellow mutineers sailed for a year, trying to find a place to settle. Unwelcome in Tubuaï, and afraid they would be captured by British forces in Tahiti, they took with them a bunch of Tahitian men and women and wandered in search of an ideal hideaway.

The mutineers landed on Pitcairn, then uninhabited, and settled there. Quickly they burned their ship, which sank in the shallow reefs of what's now called Bounty Bay, leaving no visible trace. That day they signed their life sentence, hoping never to be found by a passing vessel, least of all a Royal Navy ship.

Fast-forward to January 1957. Luis Marden, a polymath who became a senior foreign correspondent for *National Geographic* magazine, teamed with a young fellow named Tom Christian, a direct descendant of Fletcher Christian's. Marden and Christian scuba dived for the first time at Bounty Bay—and they found the remains of the *Bounty*.

In February 2013 there we were, too, diving in Bounty Bay with the descendants of the mutineers. But this time no one was trying to find a secret hideaway, nor were we seeking a sunken ship. We were looking for the real bounty of the Pitcairn Islands: its underwater ecological richness. ∎

THE BONES OF THE *BOUNTY*
Tom Christian, a direct descendant of Fletcher Christian, leader of the *Bounty* mutineers, inspects
the anchor of the infamous ship, which was burned and sunk in Bounty Bay, Pitcairn Island, in 1790.

On the one hand, I did not want to disturb the peace of whatever lay there, undisturbed, for ages; on the other hand, I wanted to explore without disturbing, and show the world what the ocean was like before we started obliterating such sacred grounds.

WESTERN PACIFIC REEFS

New Caledonia and Palau

Hot Spots of Marine Biodiversity

The night was dark, and the breeze warm. Under the cloudless sky, the ancient mariner knew exactly where home was. A small tweak on the sail, and his little dugout canoe was sailing straight to his home island in the West Pacific, 640 kilometers north of the Equator. He could guess the vague silhouette of the limestone rocks covered with lush tropical forest against the constellation Orion, now setting under the horizon.

Even before he could see his island, he knew exactly how far he was from it, for he could sense the waves breaking on the reef kilometers away. He let his right hand touch the seawater softly, as if not wanting to disturb the sleeping marine creatures. He felt the sea caressing him back.

Ancient Knowledge

In his head there was an encyclopedic amount of knowledge about his world, from the stars that guided his way among the archipelagoes, to the reefs underwater where he fished for him and his family. He, like all other elders, knew where the fish lived, where they reproduced, and how and when to catch them.

But they also knew when to leave them alone, so that there could be fish for all and forever. For generation upon generation, this traditional conservation method, the *bul,* preserved the livelihoods and food security of the people of what we know today as Palau.

However, after World War II, the traditional ways were lost to many, and westernization and population growth created increasing pressure on the marine resources of Palau, including the irony of having to provide fish for the tourists who come to observe those very fish.

Keeping Palau Pristine

Aware of his people's past and concerned about their future, Palau's President Johnson Toribiong declared at the United Nations General Assembly the first shark sanctuary in the world in 2009. In September 2013, also at the United Nations, current president Tommy Remengesau announced his intention to protect 80 percent of Palau's waters as a national marine sanctuary within the next few years, because he recognizes the advantages of a "pristine" environment for their tourism-based economy.

That is the reason why we conducted a Pristine Seas expedition to Palau. We came to support Palau's efforts to develop a sustainable future. We came to Palau to assist the government with new scientific research. We came to Palau to test how well inshore marine protected areas have performed to date, so that we can learn about their successes for the future.

And we didn't just survey the beautiful reefs of Palau: We also experienced some of the most dramatic encounters of our diving career.

We Dive In

The lagoon was calm as we motored on the German Channel, a cut blasted through the reef and dredged by the Germans at the beginning of the 20th century. The sky was a dreamy scene of cottony clouds, the type of picture urban people have on their office walls or screen savers. And then the sea boiled in front of us.

In situations when most people would get out of the water, we dive in. We adjusted our masks and snorkels and jumped in the water. Right below the surface there were tens of thousands of silvery fusiliers, swimming closely together, forming a giant ball that rotated like the shiny ball at a discotheque. We heard a bang, and the ball exploded. Where the ball was, now were two giant trevallies, decelerating as they approached us. The tens of thousands of fusiliers were nowhere to be seen. All happened in a flash. That's why these extraordinary aggregations of small fish are called bait balls. They are bait for ocean predators.

The fusiliers reappeared from below us, followed by a school of 50 black snappers, a species of fish with a massive, round head and a fierce look. We had to keep kicking our fins to keep abreast of the bait ball. The incoming tidal current was bringing water and food from the open ocean. The fusiliers and the snappers consumed that food, the microscopic organisms that form the plankton. And giant trevallies, reef sharks, rainbow runners, and dogtooth tuna showed up, hunting for the small fish.

Then two elegant ballerinas materialized as ghosts from the darkness below, moving their giant wings gracefully. The manta rays, three meters across, were also feasting on the plankton, swimming with mouths agape. They looked like giant open barrels with wings, gliding back and forth, rolling beautifully and turning around, in a rhythmic dance. I was hypnotized and could not take my eyes away from them.

Two hours passed, and the sun dived at the horizon. Around us, the bait ball was being torn apart by the predators. Another loud bang, and the bait ball became a living underwater firework. But all we could look at were the mantas, cast under their spell.

Petrie, Heart of the Coral Sea

Thirty-two hundred kilometers southeast as the albatross flies, the leaders of another country have also taken extraordinary measures to protect a natural treasure. In 2012 the governments of New Caledonia and Australia committed to create a giant protected area in the Coral Sea, the body of water that unites the two countries. They established a process to review available scientific information to determine which areas within the Coral Sea will be fully protected, and which will be multiuse areas. Pristine Seas was invited by the New Caledonian government to shed some light on some remote areas for which there were virtually no scientific studies. In other words, we were called to fill the blanks in the ocean map.

The Coral Sea is one of those places that, in spite of being close enough to a number of human settlements, is sufficiently far away so that parts of it are unexplored. Its center is sort of a "nobody's sea," an unlikely *mare incognito* in the 21st century. That's where we went in November 2013. The Waitt Foundation provided its expedition vessel, and we brought on board scientists from the Institut de Recherche pour le Développement (IRD) to survey the most remote and potentially pristine reefs off New Caledonia.

I was sitting next to Laurent Vigliola on the side of our small boat. I met Laurent, now a researcher at IRD, in the late 1990s when he was working on his doctoral thesis in Marseille, on the southern coast of France. He and I started our scientific careers in the Mediterranean, historically the most overfished sea in the world. He studied juvenile fish, and I studied what happens to the underwater ecosystem when fishing is prohibited.

As one would expect, when we don't kill fish, they take a longer time to die—and they reproduce more (see chapter 7). But the levels of recovery I observed seemed not to reach ancient abundances, for we never saw a shark in 20 years of diving in the Mediterranean. However, we were looking forward to seeing sharks in New Caledonia.

Exploring Petrie Reef

Récif Pétrie (Petrie Reef) was nothing but breakers on the surface, the sea spurting white foam to warn humans of its presence. The most remote and unknown coral reef in New Caledonia—the purest of blanks in the map—did not have land to support human habitation, and it was dangerous to navigate. Therefore, Petrie had been left alone.

I felt like an intruder breaking the last seal before entering the treasure chamber of an ancient king. On the one hand, I did not want to disturb the peace of whatever lay there, undisturbed, for ages; on the other hand, I wanted to explore without disturbing, and show the world what the ocean was like before we started obliterating such sacred grounds at a massive scale.

As we crossed the liquid boundary and entered a world forbidden to us humans, it was clear that we were entering a pristine sea. I hovered at the edge of a shallow reef, over a vertical

NOW YOU SEE ME, NOW YOU DON'T
A pygmy seahorse's body imitates the branches of the sea fan where it lives. The rich biodiversity of New Caledonia's pristine reefs can be elusive, so often do creatures blend in with their surroundings. *(page 92)*

THE ROCK ISLANDS
The Rock Islands of Palau are famous worldwide for their combination of turquoise waters and green islands.

wall that descended into the blue abyss. A handful of gray reef sharks patrolled off the wall, moving effortlessly, without any apparent movement of their tails.

I looked down. The wall was a garden of red and orange sea fans and whip corals amid hard corals that looked like ornate boulders. Like bromeliads growing on tropical trees, sea lilies the size of a dinner plate hung on the branches of large sea fans. I looked closer, and within the branches of the sea lilies I saw little yellow and brown fish, their color mirroring that of their host. Who knows how many smaller creatures were living on those fish!

Looking Closer

The closer one looks, the more life appears in front of our eyes. I looked right. A banded sea snake came out of a hole in the reef, and quickly stuck its head into another hole, looking for prey. I looked left. A family of orange clownfish swam nervously in and out of the branches of its host. Anemones sting their prey and potential predators, but clownfish have adapted to them and can swim amid them without trouble. They play hide-and-seek, eventually returning to their refuge.

I looked out again. Two coral trout—red groupers with thousands of small blue spots—were courting. The male, larger then the female, changed color in less than a second. The red body became dark, and white blotches painted a belt in its mid-body. The pectoral fins became white gloves, and white wide lips made the male look like an old silent movie star. The groupers swam closer to each other.

When they were in touching proximity, they pressed against each other and spiraled together up, toward the surface. Their swimming accelerated, and after a few seconds they split, releasing a white cloud between them—eggs and sperm, which will do their part in replenishing this wild reef where sharks rule. ▣

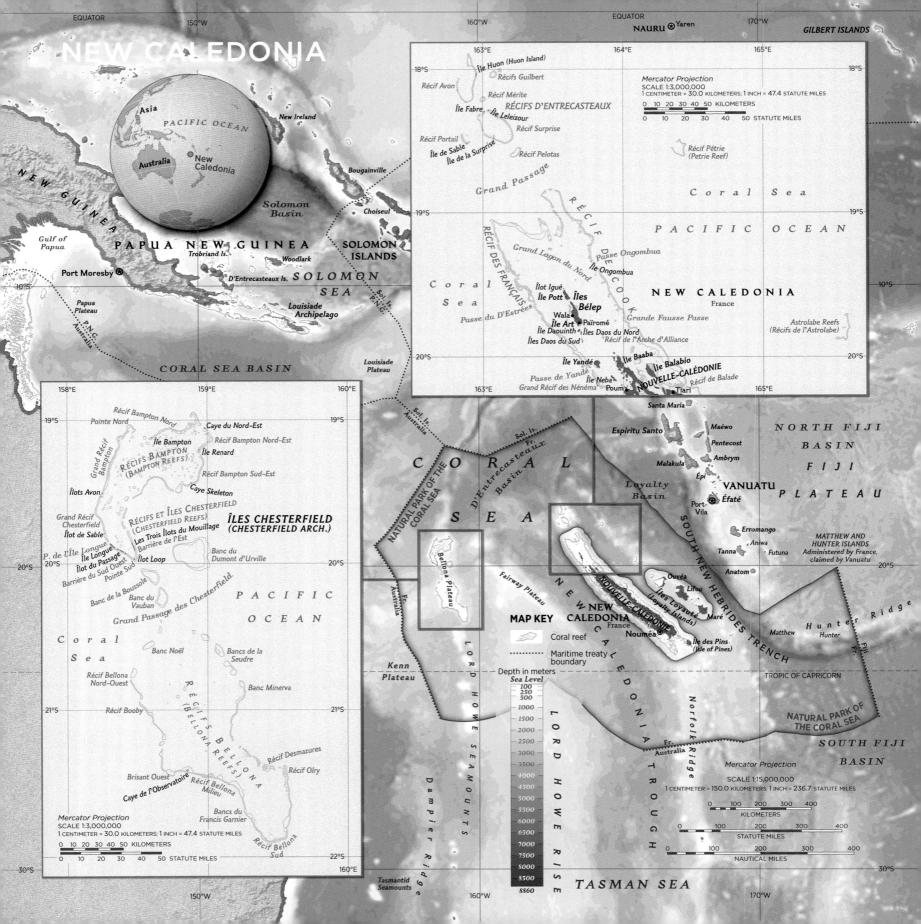

NEW CALEDONIA

Globe inset labels:
Asia
PACIFIC OCEAN
Australia
New Caledonia

Main map labels (left/lower region):

EQUATOR

NEW GUINEA
Gulf of Papua
Papua Plateau
Port Moresby ⊛
P.N.G. Australia
PAPUA NEW GUINEA
Trobriand Is.
Woodlark
D'Entrecasteaux Is.
SOLOMON SEA
Louisiade Archipelago
CORAL SEA BASIN
Louisiade Plateau
Louisiade Plateau

New Ireland
Bougainville
Choiseul
Solomon Basin
SOLOMON ISLANDS
Sol. Is. P.N.G.
Sol. Is. Australia

Upper right inset map — NEW CALEDONIA (France):

EQUATOR
NAURU ⊗ Yaren
GILBERT ISLANDS

Mercator Projection
SCALE 1:3,000,000
1 CENTIMETER = 30.0 KILOMETERS; 1 INCH = 47.4 STATUTE MILES
0 10 20 30 40 50 KILOMETERS
0 10 20 30 40 50 STATUTE MILES

Île Huon (Huon Island)
Récifs Guilbert
Récif Avon
Récif Mérite
Île Fabre — Île Leleizour
RÉCIFS D'ENTRECASTEAUX
Récif Surprise
Récif Portail
Île de Sable
Île de la Surprise
Récif Pelotas
Récif Pétrie (Petrie Reef)

Grand Passage
RÉCIF DES FRANÇAIS
RÉCIF DU COOK
Coral Sea
PACIFIC OCEAN

Grand Lagon du Nord
Passe Ongombua
Île Ongombua

Coral Sea
NEW CALEDONIA
France

Îlot Igué
Île Pott
Îles Bélep
Passe du D'Estrées
Grande Fausse Passe
Astrolabe Reefs (Récifs de l'Astrolabe)
Wala
Île Art
Païromé
Île Daouinth
Îles Daos du Nord
Îles Daos du Sud
Récif de l'Arche d'Alliance

Île Yandé
Île Baaba
Île Balabio
NOUVELLE-CALÉDONIE
Passe de Yandé
Île Neba
Grand Récif des Nénéma
Poum
Tiari
Récif de Balade

Lower left inset map — ÎLES CHESTERFIELD (CHESTERFIELD ARCH.):

Récif Bampton Nord
Pointe Nord
Caye du Nord-Est
Île Bampton
Récif Bampton Nord-Est
Île Renard
Grand Récif Bampton
RÉCIFS BAMPTON (BAMPTON REEFS)
Récif Bampton Sud-Est
Caye Skeleton
Îlots Avon
RÉCIFS ET ÎLES CHESTERFIELD (CHESTERFIELD REEFS)
ÎLES CHESTERFIELD (CHESTERFIELD ARCH.)
Grand Récif Chesterfield
Îlot de Sable
Les Trois Îlots du Mouillage
Barrière de l'Est
P. de l'Île Longue
Île Longue
Îlot du Passage
Barrière du Sud Ouest
Pointe Sud
Îlot Loop
Banc du Dumont d'Urville
Banc de la Boussole
Banc du Vauban
Grand Passage des Chesterfield
PACIFIC OCEAN
Coral Sea
Banc Noël
Bancs de la Seudre
Récif Bellona Nord-Ouest
Banc Minerva
RÉCIFS BELLONA (BELLONA REEFS)
Récif Booby
Récif Desmazures
Récif Olry
Brisant Ouest
Récif Bellona Milieu
Caye de l'Observatoire
Bancs du Francis Garnier
Récif Bellona Sud

Mercator Projection
SCALE 1:3,000,000
1 CENTIMETER = 30.0 KILOMETERS; 1 INCH = 47.4 STATUTE MILES
0 10 20 30 40 50 KILOMETERS
0 10 20 30 40 50 STATUTE MILES

Main map (center/right):

CORAL SEA
NATURAL PARK OF THE CORAL SEA
D'Entrecasteaux Basin
Bellona Plateau
Fairway Plateau
NEW CALEDONIA BASIN
NOUVELLE-CALÉDONIE
NEW CALEDONIA
France
Nouméa ⊛
Île des Pins (Isle of Pines)

Santa Maria
Espiritu Santo
Maéwo
Pentecost
Ambrym
Malakula
Épi
VANUATU
Éfaté
Port-Vila
Erromango
Aniwa
Tanna
Futuna
Anatom
Loyalty Basin
Ouvéa
Lifou
Îles Loyauté (Loyalty Islands)
Maré

NORTH FIJI BASIN
FIJI
PLATEAU
SOUTH NEW HEBRIDES TRENCH
MATTHEW AND HUNTER ISLANDS
Administered by France, claimed by Vanuatu
Matthew
Hunter
Hunter Ridge
Fr. Fiji

Sol. Is. Australia
Sol. Is. Fr.

Kenn Plateau
Fr. Australia
LORD HOWE SEAMOUNTS
LORD HOWE RISE
Dampier Ridge
Tasmantid Seamounts
TASMAN SEA

Norfolk Ridge
Fr. Australia
NORFOLK TROUGH
NATURAL PARK OF THE CORAL SEA
SOUTH FIJI BASIN
TROPIC OF CAPRICORN

MAP KEY
Coral reef
Maritime treaty boundary

Depth in meters
Sea Level
100
250
500
1000
1500
2000
2500
3000
3500
4000
4500
5000
5500
6000
6500
7000
7500
8000
8500
8860

Mercator Projection
SCALE 1:15,000,000
1 CENTIMETER = 150.0 KILOMETERS; 1 INCH = 236.7 STATUTE MILES
0 100 200 300 400 KILOMETERS
0 100 200 300 400 STATUTE MILES
0 100 200 300 400 NAUTICAL MILES

NEW CALEDONIA, a special collectivity of France, lies in the South Pacific 1,210 kilometers east of Australia. It encompasses an exclusive economic zone of 1.3 million square kilometers, which harbors a great diversity of ocean habitats, from an oceanic trench 7,919 meters deep to undersea mountains and coral reefs that are among the most species-rich areas on the planet: 25 species of mammals, 48 species of sharks, and 5 species of turtles.

In 2014 the New Caledonian government signed a decree creating the Parc Naturel de la Mer de Corail (Natural Park of the Coral Sea), which will protect critical ocean areas.

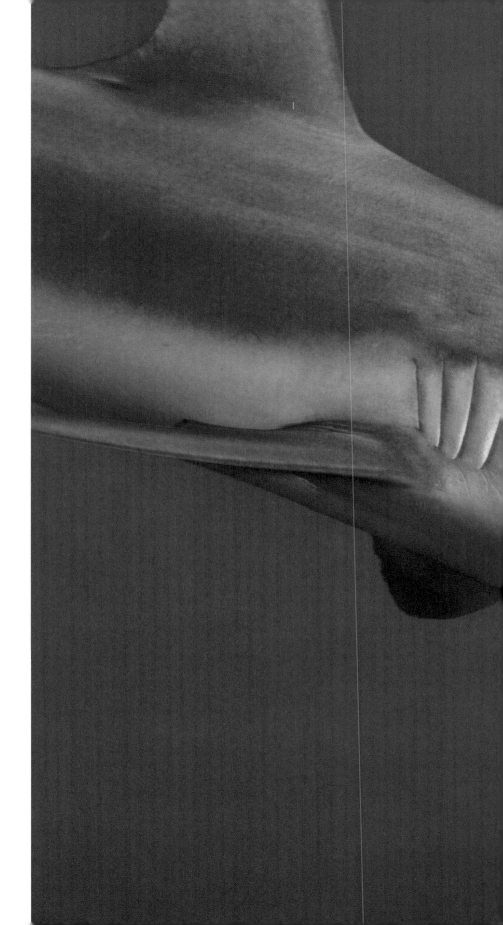

WHO'S FOLLOWING WHOM?
Although it seems like this gray reef shark
is chasing a juvenile jack, the jack is actually
taking advantage of the pressure wave
created by the shark to swim with less effort—
like dolphins swimming at the bow of a boat.

UNDERWATER FLOWERS
Two cling gobies spend their entire lives
among the tentacles of a sea lily, a close
relative of sea stars that extends its tentacles
to capture plankton in the water.

OCEAN GLIDERS
Thirty meters underwater, a large male
green turtle glides over an underwater cliff
punctuated by whip corals off New Caledonia.
(following pages)

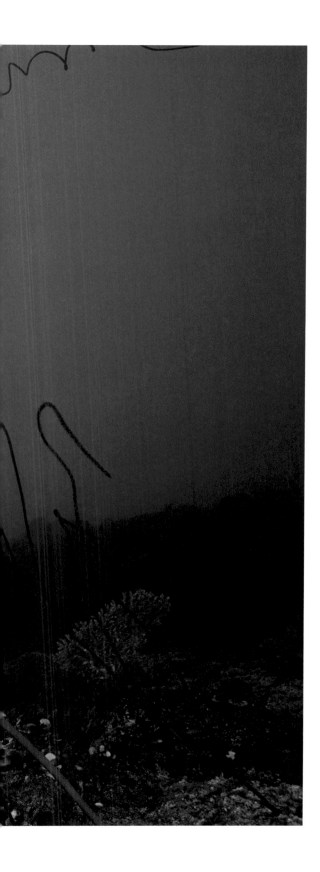

"At the same time that we are earnest to explore and learn all things, we require that all things be mysterious and unexplorable, that land and sea be indefinitely wild, unsurveyed and unfathomed by us because unfathomable. We can never have enough of nature."

—HENRY DAVID THOREAU

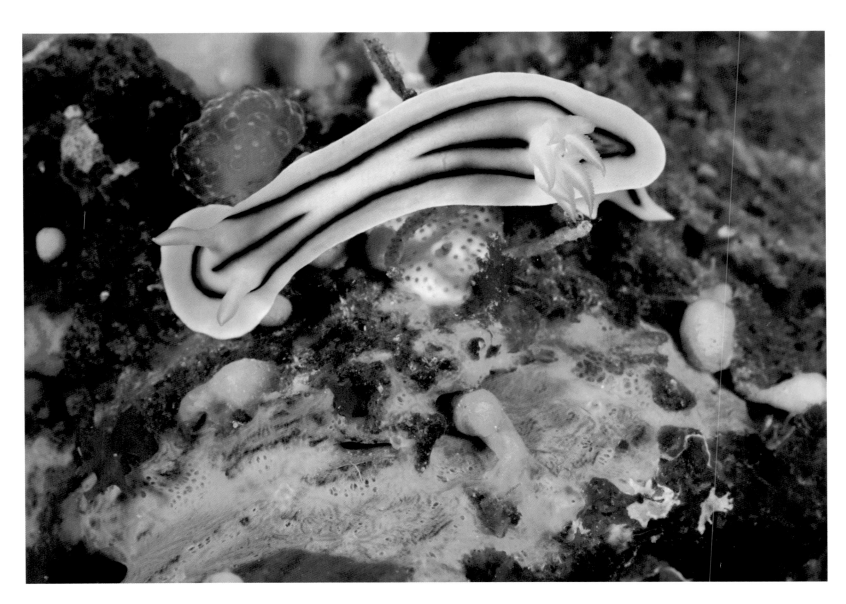

NUDIBRANCH MEETS YELLOW SPONGE, PETRIE REEF
The biodiversity of these reefs is so extraordinary that
an expert biologist can find more than 100 different
species of animals and plants in a single square meter.

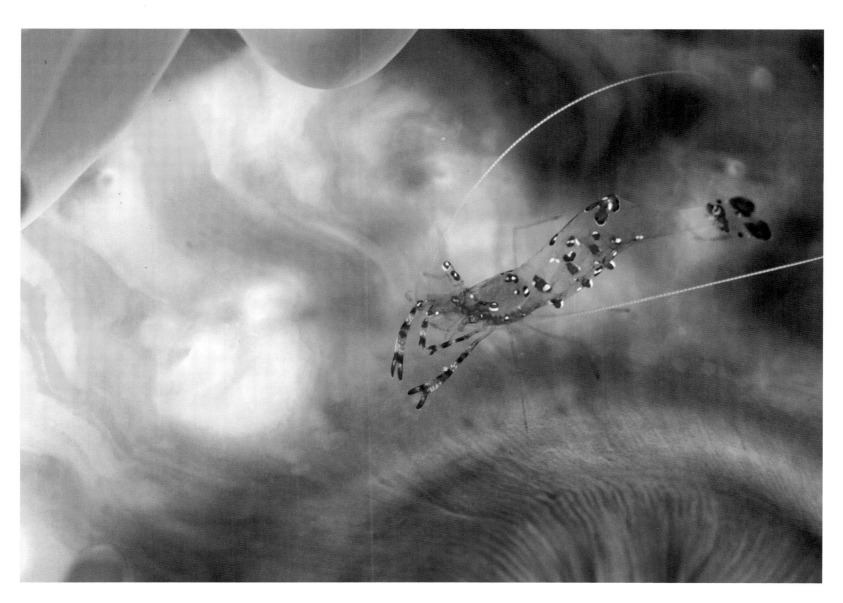

SEASCAPE IN MINIATURE
A minute shrimp lives sheltered among the tentacles of an anemone. These two species live in symbiosis: The shrimp cleans debris and parasites off the anemone, and the anemone's stinging tentacles keep the shrimp safe from predators.

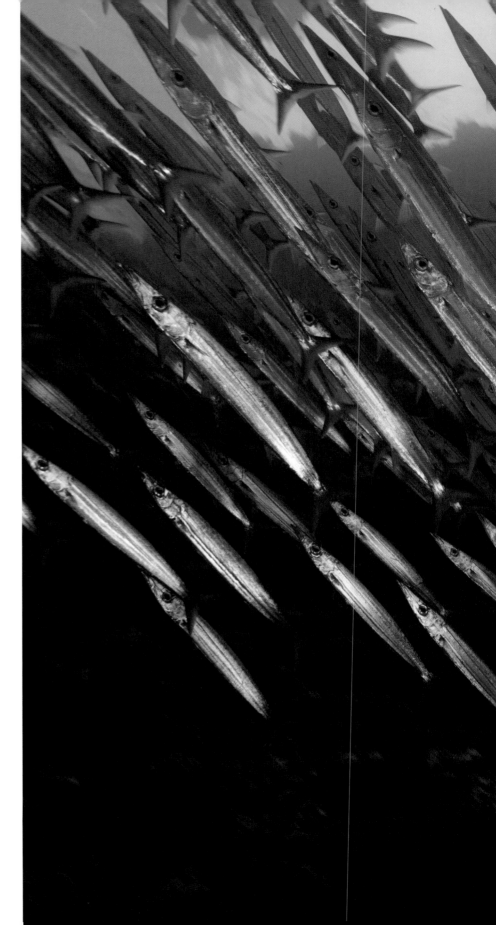

A FLEET OF SILVER TORPEDOES
Barracudas can be found in schools
of hundreds, remaining stationary or
often swimming in circles, at the pristine
Petrie Reef. The purpose of that behavior
remains unknown.

Coming Home

Before sunrise, the island was quiet. No birds flying and squawking. We turned off the engine of our small boat before reaching the shore and glided through the warm sea. Trying to be as silent as possible, we stepped onto the chilly sand, put our dry bags on the ground, and listened. I heard a faint swoosh . . . swoosh . . . As we walked closer to the source of the noise, we saw a crater on the sand. There were tracks on the sand linking the crater and the sea, the shape of a tank's tracks. Then we realized there were many more craters. The island looked as though it had been carpet bombed, and tanks had driven everywhere in between. Then I heard swoosh again.

Twenty meters ahead, something was throwing handfuls of sand out of a crater. As my eyes got accustomed to the dawn light, I saw more little sand geysers coming out of another crater, and then another. After another swoosh came some panting, and a thump, the sound of something big hopping on the sand. We tiptoed around until we reached the source of the sounds. A green turtle, three feet long, climbed out of the crater, flapping her fins as though she were swimming in water. We could sense her exhaustion: On land, the turtle is so heavy, she spends most of her energy by simply kicking sand behind her shell.

That turtle—and hundreds of others—had spent the night before laying eggs in the sand. We did not see the ritual, but we knew what had happened: First they climb from the sea up this sandy slope, up to the flatter top of the island, which rises only a few meters above sea level. Once at their chosen spot, they start digging in the sand, making those craters; they lay 100 to 200 eggs two or three feet deep in the sand, and then they work to cover them up. After their exhausting job is completed, they use their last bits of energy to crawl back to the water. Once they reach the sea they become graceful submarines again. Every green turtle will lay eggs a number of times every season, and thousands will come back within five years to Huon Island, this little sandbar in the middle of the Coral Sea, north of New Caledonia.

I followed our turtle back to the water. Her travails seemed herculean, and she stopped to rest and breathe deeply every other meter. She reached the sea as the sun was rising behind her shell. She never looked back to her nesting grounds. She put her head beneath the water and with one final push disappeared into the sea grass beds. ∎

REST AFTER A HARD NIGHT'S WORK
After spending many hours digging and laying eggs at Huon Island,
a female green turtle makes it back to the sea as the sun rises.

PALAU

Asia

PACIFIC
OCEAN

Palau

Australia

Inset map

130°E   135°E

PHILIPPINES

PROPOSED PALAU MARINE RESERVE

PALAU

FEDERATED STATES OF MICRONESIA

PROPOSED PALAU MARINE RESERVE

INDONESIA

☐ Proposed Marine Reserve
☐ Proposed Domestic Fishing Zone
⋯ EEZ

10°N
5°N
0°

0   100   200 KILOMETERS
0   100   200 STATUTE MILES

VELASCO REEF

Ngaruangl

Ngaruangl Reef

Euchelel Ngeruangl
(Ngaruangl Passage)

NGCHEANGEL
(KAYANGEL ATOLL)

Ngcheangel (Kayangle)
NGCHEANGEL (KAYANGEL ISLANDS)

Ngeriungs

Ngerebelas

Kekerel Euchel (Kayangel Passage)

Telebadel ra Ngkesol
(North Entrance)

Bkul a Chelas

NGERAEL
(NORTHWEST REEF)

Bkul a Tmederial

(West Entrance) Telebadel ra Ngerael

Toachel Ngkesol
(Kossol Passage)

NGKESOL (KOSSOL REEF)

Tochelir ra Ngebard

NGEBARD (CORMORAN REEF)

Kawasak Passage

Kloul Euchel (East Entrance)

PACIFIC

OCEAN

(Kawasak Passage) Ebil

Ngerechur

Ngerkeklau

(Ngamegei Passage) Iengel

Arukaron Point

Mengellang

AREKALONG PENINSULA

(Aiwokako Passage) Iuekakw

Ngardmau Bay

Ulimang

Bkulacheld

Ngkeklau

Alyasu Reef

Bkulangril

Toachel Mlengui

MAP KEY

Coral reef

Depth in meters
Sea Level
100
250
500
1000
1500
2000
2500
3000
3500
4000
4500
5000
5500
6000
6500
7000
7500
8000
8050

BABELDAOB (BABELTHUAP)

Namai Bay

Ochiberames (Melekeok Point)
Melekeok

Ngchesar

Idims (a Idrims Reef)

Bkurrengel

Ngchemiangel

Ngarengeivog Bay

Ngertachebeab
(Komebail Lagoon)

Arakabesan

Koror

Airai

Korak

Ngemelachel (Namelakl Passage)

Ngetngod (Goraklbad Reef)

Goraklbad Passage

NGERDILUCHES

Mutkebesang

(Malakal Harbor) Malakal

Koror (Oreor)
Ulebsechel (Auluptag)

Garreru
Toachel Mid

Uchelbeluu (Augulpelu Reef)

Ulong Channel   Ulong

Butiaur

Ngebedangel

Toachel ra Kesebekuu
(Malakal Pass)

Rael Dil

CHELBACHEB (ROCK ISLANDS)

Ngeruktabel

(Seventy Islands) Ngerukevid

Toachel Iou (Sar Passage)
Demul
Iou

Chesau (Kasao Reef)

Ongeim'l Tketau (Jellyfish Lake)

German Channel

Mecherchar

Bkul a Chesemiich
Chudel

Dmasech
Ngemelis

Sebeseb

Denges (Denges Passage)

Bailechesengel

Ngerchong

(Barnum Bay) Sngelokl

Ngercheu

Ngedbus   Olngeuaol

Kloulklubed

(Peleliu) Beliliou

PHILIPPINE

SEA

Mekaeb

(Geugel-Makaep Passage)

Bkul a Omruchel

Ngaramasch
Bkul a Medorm

Ngeaur
(Angaur)

Mercator Projection

SCALE 1:800,000
1 CENTIMETER = 8.0 KILOMETERS; 1 INCH = 12.6 STATUTE MILES

0   10   20   30
KILOMETERS

0   10   20   30
STATUTE MILES

0   10   20   30
NAUTICAL MILES

PALAU is a nation of 250 islands in Micronesia in the western Pacific. Palau's waters include most types of the world's ocean ecosystems, from a deep ocean trench to shallow coral reefs. Palau's coral reefs are famous worldwide for diverse and abundant marine life, including giant manta rays and sharks. Probably the best known natural features of Palau are its marine lakes, tucked within limestone islands covered by lush tropical forest and home to millions of stingless jellyfish. Palauans have a long tradition of managing marine resources, based on a deep knowledge of the natural history of their seas.

PALAU—PRISTINE PARADISE

The Republic of Palau claims the word
"pristine" as part of its official slogan.
Underwater, healthy coral reefs thrive,
and above spectacular limestone islands
lie covered in lush tropical jungle.

ALIENS OF THE DEEP
The nautilus resembles a squid within a beautiful chambered shell. These mollusks live in the twilight zone, below 200 meters deep. The one seen here belongs to a species found only in Palau.

LIONS ON THE REEF
A juvenile lionfish swims between patch reefs on Palau's Rock Islands. Native to Indo-Pacific reefs, lionfish are voracious eaters of small fish. Their spines contain a potent toxin and are their best defense.

GORGONIAN LANDSCAPE
Sunlight streams through the water, backlighting a large gorgonian. These colonial animals bear branches formed by thousands of small polyps, each of which consists simply of a mouth, a small digestive pouch, and eight tentacles.

SIGNS OF A VIOLENT PAST
During World War II, Palau was the site of brutal battles between Japanese and American forces. War wrecks such as this Japanese seaplane can be found throughout the lagoon. *(following pages)*

119

"We Americans, in most states at least, have not yet experienced a bear-less, eagle-less, cat-less, wolf-less woods. Germany strove for maximum yields of both timber and game and got neither."

—ALDO LEOPOLD

BREAKING UP THE PARTY
A few seconds earlier, this school of fusiliers swam in a tight school—a "bait ball"—that exploded once the giant trevallies dove in after them.

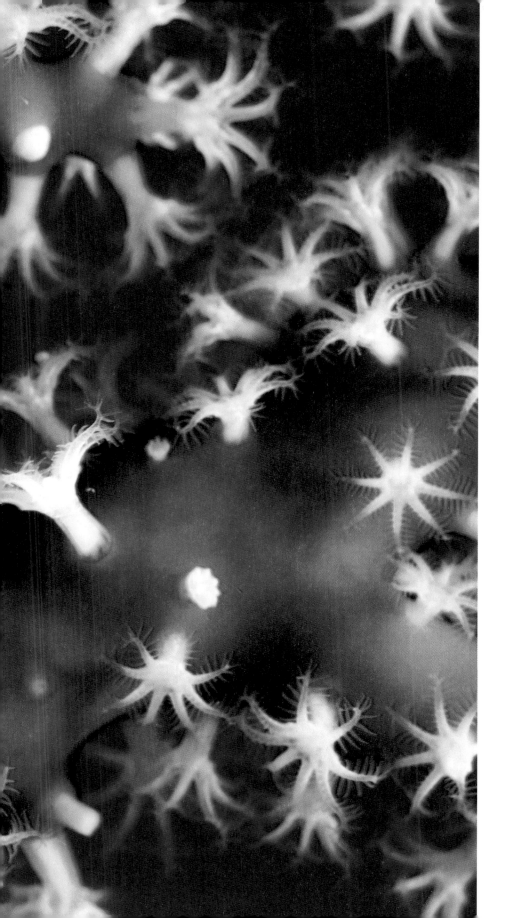

EYES WIDE OPEN
A small goby hides among the branches of a gorgonian. Orange and white markings on an otherwise translucent body allow this little fish to blend in almost seamlessly against the branches of its brightly colored surroundings.

MOUTHS WIDE OPEN
Two and a half meters across each, these giant manta rays swim with mouths agape at German Channel, filtering plankton out of the seawater. Before high tide, plankton accumulates on reef channels.

Jellyfish Lake

As incredible as it may sound, the greatest tourist attraction of Palau is a lake with five million jellyfish. Popularly called Jellyfish Lake, it is a secluded body of water on Mecherchar, one of hundreds of limestone islands covered by lush tropical vegetation in this island nation. Tourists come from all over the world, especially from Asia, not only to see the jellyfish but also to swim among them.

We motored on our fast boat through a maze of marine channels amid the Rock Islands early in the morning, to beat the tourist groups. The water ranged from turquoise to navy blue, and it was so clear that more than once I could see the coral reef so clearly beneath us that I thought we were going to run into it. We arrived at a little dock on one of the islands, where two sleepy rangers inside a wooden park entrance checked our permits. After the formalities, we hiked up and down a hill, and ten minutes later there it was: an enclosed lake, surrounded by a thick forest. It is considered a marine lake because the water in it is seawater that seeps in through underwater channels, even though the lake is completely surrounded by rock and jungle.

I had been waiting for this moment for 25 years, ever since I saw a photo in *National Geographic* magazine, taken by David Doubilet, of a free diver submerged in green water and surrounded by hundreds of jellyfish. Now here I was, at the same place, ready to have that same experience. I glided into the warm salt water, and it felt as if I was floating in a primordial universe. There were jellyfish all around me, ranging in size from a thimble to a watermelon. They moved by pulsating in all directions, slowly but surely. Countless brown hearts beating in unison surrounded me. I wanted my heart to beat alongside them. I have visited some of the most remote places in the ocean, full of sharks and huge fish schools, but these little animals, virtually water inside a living bag, made me feel closer to nature than anything ever had before.

Palau is doing it right. Only one such lake is open to the public, who pay an entrance fee. Rangers control access to the lake and enforce the regulations. And the entire archipelago around the lake is untouched. No houses, no hotels, no human signs. This is why people leave their day-to-day lives and fly here: to experience undisturbed nature. ∎

THE PRIMORDIAL SEA
Isolated from the sea but for narrow underground channels, Palau's famous Jellyfish Lake is dominated by *Mastigias* jellyfish, transporting us to a seascape similar to what the Precambrian ocean might have looked like more than 500 million years ago.

I found it incredible that we were the first humans down there. Within the darkness that surrounded us, I felt like an astronaut landing on a mysterious planet.

EASTERN PACIFIC ISLANDS

Desventuradas and Cocos Islands

Remote Ocean Outposts

What kid has not dreamed of exploring some place no other human has seen before? What scientist has not dreamed of discovering a species never observed before? I was one of those kids, and then I became one of those scientists. I am a fortunate dreamer, too, because desiring those dreams so intensely made me realize them. Pristine Seas has taken our team to some unexplored places, and we have discovered species new to science—repeated times. And we have more places on our list to explore in the next few years. Once I was asked if I ever get tired of exploring. My answer was a robust "Never!" I have kept that child's curiosity strong within me.

Every time we are about to enter one of those secret natural underwater halls, I have the same feelings as the first time. It's more than excitement; it's like that unbearable passion before the first, long-desired kiss. No discovery epitomizes those sensations better than our deep-sea exploration of the Desventuradas, two small islands about 500 nautical miles off the northern coast of Chile.

Diving the Desventuradas

It was February 25, 2013. The *Argo,* our ship, was stationed above an undersea mountain whose top stopped 15 meters short of the surface. A few kilometers away was San Félix, an island so small it's invisible on a satellite photo of the eastern Pacific. Yet to us, that day, San Félix was our whole world.

"Enric, are you ready?" Avi Klapfer called to me. "I cannot wait!" I replied.

Avi is a former Israeli Navy officer who sailed the Pacific with his wife for years, ending up in Costa Rica, where he owns a diving center that operates mostly at Parque Nacional Isla del Coco (Cocos Island National Park). The *Argo* and the *DeepSee* submersible it carries are his flagships. Avi is a slim man with unending enthusiasm yet Zen-like calmness, a permanent smile, and an extraordinary knowledge of the sea, diving, and engineering. He and his partners designed their submersible from scratch, merging a combined hundred years of experience into the perfect vessel for exploring the underwater twilight zone. After a successful expedition to Cocos and some unexplored seamounts nearby in 2009, we decided to charter Avi's ship and submersible again. What I love the most about them is their can-do attitude. To them, problems are only challenges to be solved.

I stepped into the submersible, secured to the ship by several strong lines as if caught inside a spiderweb. Careful to avoid stepping on any of the delicate gauges on board, I took my seat forward, while Avi sat behind me. Alex Muñoz—my great friend and conservation partner, who leads Oceana in Latin America—followed and took his seat next to me. Avi made a sign to his crew, and they lowered the acrylic dome on top of us, latched it, and gave Avi the OK sign. We were sitting inside a transparent sphere full of electronics, inserted in a yellow frame with electrical engines and propellers—and, most important, a high-definition camera in a housing rated to go down to 500 meters in depth that we could manipulate from inside the sub.

"Oxygen, 20.8, 20.8 . . . Battery, 95 percent, 95 percent . . ." Avi said, slowly and clearly, going through the checklist to test all systems on board.

The ship support team gave us authorization to dive. The crew towed us off our berth on the stern of the *Argo*. We bobbled on the surface, while juvenile masked boobies, comical-looking seabirds, aggregated around the sub, dipping their heads in the water to check us out. They were like a bunch of teenagers gathering around a flashy new car that just arrived in town.

We released air, creating a noisy column of bubbles, and started the descent. Alex and I looked at each other, smiled, and shook hands and arms warmly. We were the first humans to dive along the slopes of that undersea mountain.

Down Steep Slopes

Our diving protocol consisted of diving as deep as possible and then ascending slowly along the slope, stopping to catalog, photograph, and film marine life and geologic formations. A normal dive would take between two and three hours, depending on how much life we encountered along the way. That seemed an impossible limit to us, though: It is very difficult to move down a rocky slope full of animals we have never seen before without stopping again and again. Therefore, we made a promise to ourselves to be very disciplined and avoid stopping

every other minute on our way down—unless, of course, we observed something extraordinary . . . but everything looked so extraordinary to me!

Because the refraction of light on the acrylic dome is similar to seawater's, we had the impression of being in the water with no dome around us. The first hundred meters we were visited by many fish, from a school of yellowtail jacks that orbited us at high speed for 20 seconds to sea bream–like morwongs so curious that they touched the sub sphere as though they were kissing it. At 90 meters we observed giant Juan Fernández lobsters moving slowly over a large rock. A few days before we had taken one on board for a brief examination, and it weighed 17 pounds.

The volcanic bottom was dark and populated with whip corals, like giant springs planted on the rock, and white sea fans, each perhaps hundreds of years old. At 140 meters, a couple dozen bright dots moved randomly over the bottom. I asked Avi to get closer, and he maneuvered the sub smoothly and quietly. Whatever they were, they vanished into the crevices of the rock before we could see them clearly. Looking closer into one of the crevices, we glimpsed a yellow patch surrounding what looked like an eye. Avi switched our lights off, and we waited a few minutes. The ambient light was attenuated because of the depth, but it was sufficient for us to see that little fish were coming out of those crevices. After five minutes Avi switched on the lights again, and we saw about 50 fish, each the size of a small lemon, with cute round heads and fins, bright orange and yellow. The fish darted back to the crevices, escaping from the bright lights. I had never seen that species before; hence we

needed to take good footage so that we could
identify it afterward.

Lobster Discoveries

At 200 meters we reached a large ledge where a
couple dogfish were swimming over a field of sea
urchins with spines as thick as pencils. The slope
became steeper, now 45 degrees. I felt like a kid
sledding down a winter mountain. We needed to
speed our descent because we already had spent
one hour underwater. We were flying over forests
of sea fans and sea whips, following the contours
of the bottom. We noted volcanic ridges that
looked like underwater Hadrian's walls, knowing
we would use them as references so that we could
relocate some of the wonders we had spotted on
the way down.

At 290 meters, we saw a dozen spiny lobsters
clustered together. Two hundred ninety meters?
That was far below the known range of the Juan
Fernández lobster. I asked Klapfer to steer the sub
toward them, and he obliged me, kindly as always.
As we got close to the lobsters, I realized that these
were not Juan Fernández lobsters. First, they were
smaller, only 25 centimeters long, including the
antennae. Second, they were more delicate, with
long, elegant feet. The edges of their abdominal
plates were spiny. I remembered reading a paper
mentioning a species of dwarf lobster caught occasionally in deep traps intended to catch the giant
lobsters in the Archipiélago Juan Fernández. That
was it. These were dwarf lobsters.

Because only a few fishers ever come to the
Desventuradas to catch Juan Fernández lobsters,
and because they stay in shallower waters, this was
probably an unfished lobster population. Therefore,
we had a rare opportunity to study a population of
a species of commercial interest that likely had
never been touched. Those lobsters could provide a
baseline for a pristine population of dwarf lobsters.

Recognizing the significance of this discovery,
we decided to count the dwarf lobsters in a

HIDING IN PLAIN SIGHT
A frogfish rests completely
immobile, looking like a sea
sponge, off Isla Manuelita
at Cocos Island National
Park. Frogfish are ambush
predators, nabbing the
small fish that swim in front
of them in a fraction of a
second. *(page 130)*

THE UNFORTUNATE ISLANDS
The Desventuradas (meaning
"unfortunate" in Spanish)—
San Ambrosio (left) and San
Félix (right)—are remote and
uninhabited but for a small
outpost of the Chilean Navy.

standardized way. We drove the submarine straight down for one minute, following the steepest slope. We noted our initial and our final depths and estimated the steepness of the slope, which by simple trigonometry gave us an estimate of the length of that transect. We counted all lobsters within four meters on each side of the sub. We estimated their size using two laser pointers attached to our underwater camera. Placed in parallel, they represented exactly one foot in distance. Applying simple arithmetic, we estimated we were observing 185 pounds of dwarf lobster per hectare. Thus we established a baseline for a pristine dwarf lobster population—a much needed reference point for the study of exploited populations elsewhere.

After three hours underwater we reached a depth of 400 meters. We stopped the sub in front of a ledge where orange sea fans were growing upside down from an overhang.

"So, should we start going up, or you want to continue going down?" said Avi with a wink. We could not go down much farther, since the maximum depth for the sub is 450 meters, and we had already used much of the battery power during the last three hours. But it was a nice thought.

"Nah . . . let's go up, slowly! So we can film all those creatures we left behind on our way down," I replied. "But let's wait and enjoy the peace down here for a few minutes."

And so we did, quietly, everyone absorbed in their own feelings. I found it incredible that we were the first humans down there. Within the darkness that surrounded us, I felt like an astronaut landing on a mysterious planet.

On the way back to the surface we stopped along volcanic ridges, peeked into dark caves, cataloged dozens of species, and discovered two new species of fish on that dive alone—including the yellow and orange fish that had escaped from our lights on our way down, now known as *el chilito*. But most important, we populated with new knowledge one pixel in a vastly unknown ocean. ▪

DESVENTURADAS ISLANDS

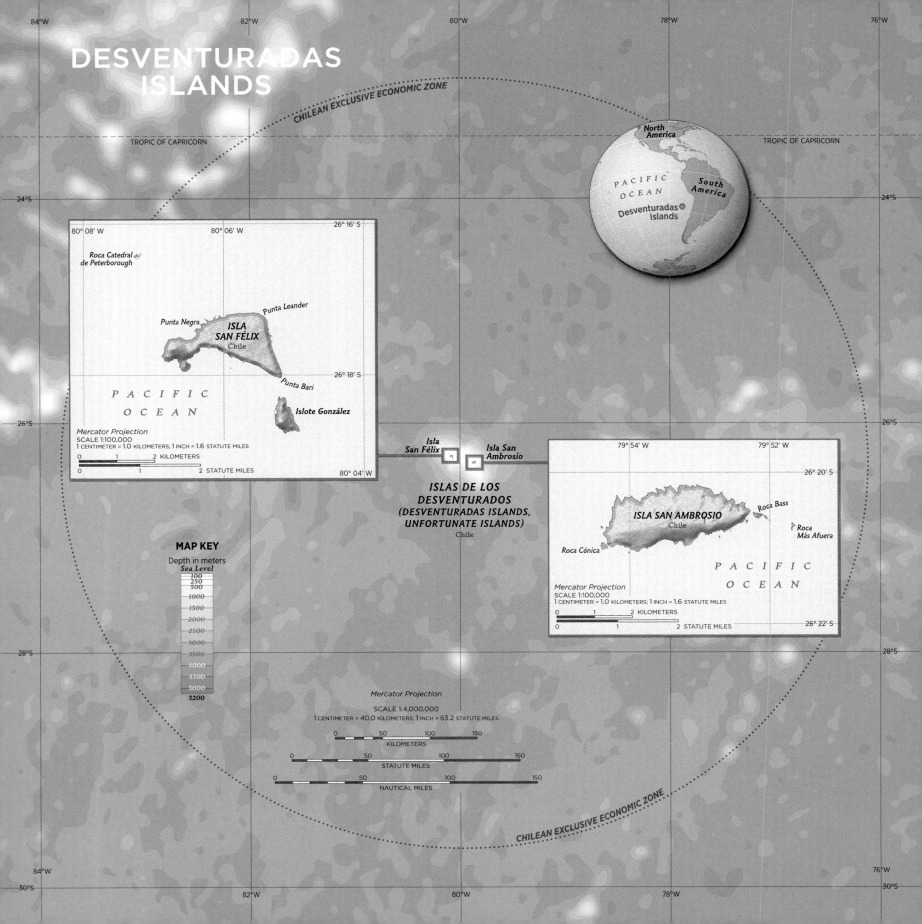

CHILEAN EXCLUSIVE ECONOMIC ZONE

TROPIC OF CAPRICORN

North America

PACIFIC OCEAN

South America

Desventuradas Islands

TROPIC OF CAPRICORN

80° 08' W
80° 06' W
26° 16' S

Roca Catedral de Peterborough

Punta Leander

Punta Negra

ISLA SAN FÉLIX
Chile

26° 18' S

Punta Bari

PACIFIC OCEAN

Islote González

Mercator Projection
SCALE 1:100,000
1 CENTIMETER = 1.0 KILOMETERS; 1 INCH = 1.6 STATUTE MILES

0   1   2 KILOMETERS

0   1   2 STATUTE MILES

80° 04' W

Isla San Félix

Isla San Ambrosio

**ISLAS DE LOS DESVENTURADOS
(DESVENTURADAS ISLANDS,
UNFORTUNATE ISLANDS)**
Chile

79° 54' W
79° 52' W

26° 20' S

ISLA SAN AMBROSIO
Chile

Roca Bass

Roca Más Afuera

Roca Cónica

PACIFIC OCEAN

Mercator Projection
SCALE 1:100,000
1 CENTIMETER = 1.0 KILOMETERS; 1 INCH = 1.6 STATUTE MILES

0   1   2 KILOMETERS

0   1   2 STATUTE MILES

26° 22' S

MAP KEY

Depth in meters
Sea Level
100
250
500
1000
1500
2000
2500
3000
3500
4000
4500
5000
5200

Mercator Projection

SCALE 1:4,000,000
1 CENTIMETER = 40.0 KILOMETERS; 1 INCH = 63.2 STATUTE MILES

0   50   100   150
KILOMETERS

0   50   100   150
STATUTE MILES

0   50   100   150
NAUTICAL MILES

CHILEAN EXCLUSIVE ECONOMIC ZONE

THE DESVENTURADAS are two small islands, San Ambrosio and San Félix, plus two islets 850 kilometers off the northern coast of Chile. They are uninhabited except for a small outpost of the Chilean Navy on San Félix. The islands are the only emergent part of a volcanic ridge that rises from the seafloor 4,000 meters deep. Representing a zone of transition between subtropical and subantarctic waters, the fauna here include subtropical species (such as long spine sea urchins) and temperate species (such as the Juan Fernández fur seal). Most of the area of the Desventuradas are deep-sea habitats, including more than 20 undersea mountains, home to many animals found nowhere else on Earth.

SCHOOL OF SCALLOPED HAMMERHEADS
Scalloped hammerheads are one of the attractions of Cocos Island National Park. The Bajo Alcyone, in particular, is one of the few places in the world where it is possible to dive with more than 200 of them.

QUICK ESCAPE
A peacock flounder exhibits its flashy colors as it speeds away from a diver. Typically flounders are difficult to see, as they bury themselves in the sand and hide before ambushing their prey.

ONE AMONG MANY
A bottlenose dolphin chases a school of bigeye trevallies off Cocos Island. Trevallies are too large for the dolphin to eat, but this chase forced one stressed trevally to vomit its stomach contents—which the dolphin then ate. *(following pages)*

"There is some quality in man which makes him people the ocean with monsters and one wonders whether they are there or not. In one sense they are, for we continue to see them . . . Men really do need sea-monsters in their personal oceans."

—JOHN STEINBECK

A SCHOOL OF BLUE AND GOLD
Native to the eastern tropical Pacific, the blue and gold snapper typically schools amid shallow rocky reefs and eats small crustaceans such as crabs and shrimp.

IRIDESCENT BIGEYES
The bigeye catalufa grows to 30 centimeters long and inhabits
deeper rocky reefs off Cocos Island. This school was photographed
at Bajo Alcyone, a seamount that peaks at a depth of 30 meters.

A LOVING EMBRACE
Green turtles aggregate to mate at Cocos Island and then swim hundreds of kilometers to mainland Costa Rica, where they will lay their eggs on sandy beaches.

A River
of Sharks

The *Argo* was anchored at night near an island 300 nautical miles from the shores of Central America. There were no lights except for those of our own ship. We had dived under darkness hundreds of times before, but that night we did it with more trepidation than I can ever remember.

At first, nothing seemed different—just another night dive in the tropics. Very few fish are to be seen, and invertebrates such as sea stars and sea urchins come out of their shelters to eat, unmolested by daytime predators. Our handheld flashlights drew circles of color and texture on the black canvas as we moved them slowly back and forth. One lonely whitetip reef shark swam indifferently in front of our lights, and shortly after it disappeared into the darkness. We swam along the bottom, and seconds later saw another whitetip reef shark, and then another. One minute later we witnessed the formation of what can only be called a shark pack, with a dozen of them swimming in unison, calmly and slowly.

All of a sudden one of the sharks darted out of the group and disappeared. The other sharks followed immediately. We moved our lights to where the sharks were going and saw a red squirrelfish coming toward us, fast, like a mini-torpedo. Behind it came the sharks, swimming frantically. I moved aside to avoid being bumped by them. I lost sight of the sharks for a second, but Manu San Félix, our underwater cinematographer, who loves sharks more than any other marine animal, flashed his light to me. I swam toward him as quickly as I could. Before I reached him, I heard a sound like thick sandpaper filing on a rock. I pointed my light toward the noise. About 20 sharks were trying to stick their heads and bodies into a crevice in the reef 30 centimeters wide. Their bodies were vibrating and shaking feverishly, as though they were possessed. The commotion increased as more sharks materialized from the darkness and joined in the fray. The tension grew until we heard a loud crack. So many sharks pushing had broken the coral head, beneath which a fish was hiding. One of the sharks got lucky and swam away from the group with half a red squirrelfish sticking out of its mouth. Another shark shot out toward it, and a battle for half a fish ensued. Five seconds later, the fight finished, as the two lucky sharks got their share of the prey. But the rest of the pack resumed swimming, calmly continuing their search for prey. ■

THE NIGHT HUNT
Whitetip reef sharks congregate every night at Isla Manuelita, Cocos Island National Park, to hunt. They swim in tight formation and fight over every fish they catch.

COCOS ISLAND

North America

Cocos Island

South America

PACIFIC OCEAN

GUATEMALA BASIN

West Cocos Seamount

Las Gemelas Seamounts

SEAMOUNTS MARINE MANAGEMENT AREA

Isla del Coco (Cocos Island) Costa Rica

COCOS ISLAND NATIONAL PARK

COCOS

RIDGE

Ciudad Cortes

Bahía de Coronado   **COSTA RICA**

Isla del Caño

Punta Llorona

Península de Osa

Punta Salsipuedes

Cabo Matapalo   Punta Banco

PANAMA

Golfito

La Concepción   David

Golfo Dulce

Puerto Armuelles

Península Burica   Bahía de Charco Azul

Punta Burica

Costa Rica / Panama

PANAMA BASIN

PANAMA FRACTURE ZONE

MAP KEY

- - - - Maritime treaty boundary

Depth in meters
Sea Level

100
250
500
1000
1500
2000
2500
3000
3500
4000
4500
4600

Medina Seamount

Costa Rica / Ecuador

Mercator Projection
SCALE 1:3,000,000
1 CENTIMETER = 30.0 KILOMETERS; 1 INCH = 47.4 STATUTE MILES

0 25 50 75 100
KILOMETERS

0 25 50 75 100
STATUTE MILES

0 25 50 75 100
NAUTICAL MILES

Inset map — Isla del Coco

87° 06' W   87° 04' W   87° 02' W

5° 34' N

PACIFIC OCEAN

Estrecho Chale

Punta Agujas

Isla Manuelita

Punta Quirós

Isla Pájara

Bahía Weston

Bahía Chatham

Punta Pacheco

Península Presidio

Isla Cáscara

Isla Cónico

Punta Presidio

Punta Ulloa

Isla Piedra Sucia

Punta Gissler

Bahía Wafer

Cabo Barreto

5° 32' N

Punta Don Felipe
Punta María

Cabo Atrevida

Punta El Coronel

ISLA DEL COCO (COCOS ISLAND)
Costa Rica

Cabo Lionel

Isla Montagne

4° N

Cabo Descubierta

Bahía Inútil

Islas dos Amigos

Bahía Yglesias   Isla Juan Bautista

Punta Montealegre

Roca Sumergida   Bajo Alcyone

Punta Sofía

Isla Rafael (Meule Island)

Punta Turrialba

Cabo Dampier

5° 30' N

Panama / Colombia

Mercator Projection
SCALE 1:100,000
1 CENTIMETER = 1.0 KILOMETERS; 1 INCH = 1.6 STATUTE MILES

0 1 2 3 KILOMETERS

0 1 2 3 STATUTE MILES

ISLA DEL COCO (COCOS ISLAND) is Costa Rica's marine jewel. Located 550 kilometers off its western coastline, this Pacific island and 1,000 square kilometers around it were protected as a national park in 1978 and designated as a World Heritage site in 1997. The waters of Cocos are very productive and contain one of the world's densest concentrations of large marine predators such as sharks. Sixty-five kilometers south of Cocos runs an undersea mountain chain called Las Gemelas—"the twins." Covered by deep-sea corals, it harbors many species of invertebrates and fish yet to be described by scientists.

UNDERSEA SCAVENGERS
A group of wrasses gathers around a dead crab below the kelp canopy at San Ambrosio. At first they bit the lifeless body timidly, which attracted other fish that came and devoured the whole crab in a minute.

SCHOOLING YELLOWTAIL JACKS
Atop a seamount north of San Félix, the Pristine
Seas team members found themselves surrounded
by nearly 500 of these 1.5-meter-long yellowtail jacks.
These fish frequent the tops of seamounts, productive
environments that attract schools of prey fish such as
sardines and anchovies.

CAT FIGHT
Near San Ambrosio, two moray eels stand off on barren ground—rocks stripped of kelp by long-spine sea urchins. Behind the morays lingers a large sea star, natural predator of the sea urchins.

THE LONG COMMUTE HOME
Large schools of fish migrate from the shallows to a seamount off San Ambrosio every day, covering the top of the rock and eating the plankton that aggregates around it, unfazed by visiting divers. *(following pages)*

157

"I would feel more optimistic about a bright future for man if he spent less time proving that he can outwit Nature and more time tasting her sweetness and respecting her seniority."

—E. B. WHITE

LOCAL GIANTS
The Juan Fernández lobster, a species found exclusively at the Juan Fernández and the Desventuradas Islands, can reach a weight of more than 8 kilograms—17 pounds. These creatures are found abundantly at depths of 100 to 200 meters.

SOLDIERS IN FORMATION
Soldierfish feed on shrimp and crabs camouflaged within the caves and crevices at San Ambrosio.

Back From
the Dead

The top predator at the Desventuradas Islands is not the typical reef shark, or a grouper with a mouth big enough to swallow a diver. It is not a fearsome animal that kills at night, either. The largest predator here is the Juan Fernández fur seal *(Arctocephalus philippii),* the cutest carnivore we have found in any of our Pristine Seas expeditions to date.

The Juan Fernández fur seal is actually a sea lion. These creatures spend much of the day hanging out on rocky platforms near the water. As the *Argo* approached the islands, we saw five of these adorable animals. When we approached them, it was like someone brought free candy to a school. The sea lions raised their heads, became very excited, and dragged their fat bellies from rock to rock until they jumped in the water.

Underwater, the sea lions became torpedoes of enormous grace and elegance. Their eyes were as large as a Japanese cartoon character's, and their looks pierced us as they swam very fast between us divers. After playing with our bubbles and checking us out very closely, they just hung out with us, their tails on the surface and their heads hanging down like bats.

The Juan Fernández sea lion lives only in the Desventuradas and the Archipiélago Juan Fernández (Juan Fernández Archipelago), 800 kilometers farther south. It was an abundant species before European whalers and hunters started to exterminate them, hunting them for their skins and oil. In the Juan Fernández Archipelago alone, between three and five million were killed between the 17th and 19th centuries. By 1880 the scientific community declared them extinct. Fortunately, a few individuals survived and were able to start replenishing their population. In 1970, about a hundred years after having been seen for the last time, two juvenile sea lions were observed at San Ambrosio, one of the Desventuradas. In 1975 300 individuals were observed.

In February 2013 we saw only five sea lions at San Ambrosio. The strong winds restricted our work to the northern side of the island, and the Chilean Navy did not allow us to dive around San Félix, where there is a military base. Therefore, we cannot determine how many sea lions are left altogether in the Desventuradas. I only hope that the sea lions are coming back, and that the restricted access will keep them safe. ■

ONCE THOUGHT EXTINCT
Technically a sea lion, the Juan Fernández fur seal escaped extinction.
Five individuals were observed during the recent Pristine Seas expedition.

While one considers some new adventure, the universe seems to idle; but the moment one truly commits to an idea, cosmic gears must be set in motion, because opportunities arise and the unthinkable becomes possible.

WESTERN AFRICA'S COASTLINE

Gabon

Where Hippos Meet Whales

December 18, 2000. A skinny white man and a group of Baka Pygmies emerge from the thick jungle and step onto a sandy beach. The white man smiles proudly. The pygmies cannot believe what they see. Having spent their entire lives under a permanent canopy, their minds cannot comprehend the vastness of the gigantic river in front of them. They drop their belongings and approach the water hesitantly. After 456 days on foot across the wildest jungles of Congo and Gabon, National Geographic Explorer-in-Residence Mike Fay and his support team have arrived at the remote beaches of Gabon and are now facing the Atlantic Ocean. But they aren't the only ones there.

As wild a beach as any in Africa, Gabon's Petit-Loango Reserve encompasses swamps, savanna, rain forest, and Atlantic beaches. Left undisturbed by humans, hippopotamuses venture into the surf to mate, play, or merely commute from lagoon to savanna, getting a lift from the buoying water rather than toiling on land.

As I read David Quammen's account of Mike Fay's Megatransect in the August 2001 issue of *National Geographic,* the little kid in me started to wander in dreamland. Surfing hippos! I had never heard of that. Would that be possible? A few pages later, boom, a photo of a big fat hippo bodysurfing.

I added "See surfing hippos in Gabon" to my bucket list. No need to add Mike Fay to my hero list; he was already there.

In 2012 Mike Fay was not just a hero to me; he had become a friend. He had joined a Pristine Seas expedition, and a great collaboration had developed. We only saw each other for short periods of time—he was living between Alaska and Gabon and I was based in Washington, D.C.—but when we were together I felt energized by his drive for conserving our precious nature and his take-no-prisoners approach. During one of his visits to Washington, he urged me to organize a Pristine Seas expedition to Gabon. He had been talking to President Ali Bongo Ondimba about the need to expand Gabon's national park system into the sea. But little was known about Gabon's underwater world, and an expedition could inspire the president.

While one considers some new adventure, the universe seems to idle; but the moment one truly commits to an idea, cosmic gears must be set in motion, because opportunities arise and the unthinkable becomes possible. After my chat with Mike, my friend Ted Waitt asked me if I would be interested in an expedition to Gabon using the Waitt Foundation's vessel. The timing could not have been more perfect. In October 2012 we left the port of Libreville to sail our own marine megatransect. We had three weeks to explore about 800 kilometers of coast, from the border of Congo in the south to the border of

Equatorial Guinea in the north. That probably would not give us enough time to wait for surfing hippos, but it would allow us to catch the first glimpse of an unexplored underwater world.

Making Our Way Through Mud

Gabon's coast is a never ending beach, dotted with mangroves and coastal lagoons. We were off Parc National de Mayumba (Mayumba National Park), on the southern shores of Gabon, between the Ogooué and the Congo Rivers, two leviathans that discharge enough water in one minute to flood Central Park under a meter of water. The water carries mud, sand, and billions of tiny particles that result from the decomposition of leaves, trunks, and animals upstream. When it discharges into the Atlantic Ocean, it turns the coastal waters of Gabon into a dense green soup above and a vast sandy plain below. Our team is used to diving on coral and rocky reefs with clear waters; here we felt totally out of our element.

Diving in mud is not the most exciting experience. We knew that there is more diversity and abundance of marine life where the sea has a hard bottom, and therefore we looked for rocky oases in the middle of the muddy desert. We had studied a French geologic chart produced for oil exploration and identified some deep rocky reefs off Mayumba. We'd start there.

We jumped in the water and looked down. We could not even see our fins. We did not feel comfortable diving down in zero visibility; there could be old fishing nets or lines on the bottom, invisible yet treacherous to swim into. Therefore we decided to explore first with Waitt's underwater robot—a remotely operated vehicle, or ROV, equipped with an underwater camera that sends live images back to the vessel through a tether.

We sent the ROV down. For the first ten meters the water was so murky that we only saw green. At 15 meters depth, the water cleared up a bit, but it was dark because of the dense cloud above that blocked the sunlight. River water stays on top because freshwater is less dense than seawater; as long as we swam below that upper layer, we should be fine. At 40 meters, the ROV hit the bottom. We moved the ROV along horizontally. For half an hour we saw nothing but sand with ripple marks. Then a patch of what looked like plant detritus appeared on the screen, and finally the rocky reef we were looking for, about 80 meters off the stern of our vessel. We decided to dive on that rock.

We jumped into the pea soup and descended, following the cable of the ROV, which we left stationed on the reef, our Ariadne's thread. The reef was a rocky slab only 1.5 meters tall, but that provided enough relief and overhangs to support red algae, orange cup corals, red sea fans, and whip corals. Among them we saw small groupers, snappers, and wrasses. Manu San Félix was next to me, filming the reef, when, all of sudden, I saw a shadow behind him: a shadow so large it could have swallowed him. Manu turned around and aimed his lights at whatever it was. I heard a bang, low and loud. The shadow disappeared. Manu looked at me and signaled "Huge grouper" in our underwater sign language. It was a goliath grouper, a critically endangered beast of a fish that can grow up to 2.4 meters long, with a top recorded weight of 358 kilograms. The goliath grouper was the top predator in Atlantic tropical reefs, but fishing has all but eradicated it. It was a great sign that the species was still present in Gabon.

A week later we did a morning dive west of Parc National de Pongara (Pongara National Park), near the capital city of Libreville. We swam down the anchor line, and the water was clearer than anywhere else in Gabon; we were all very happy. I was watching Kike Ballesteros and Manu San Félix diving below me, and Kike disappeared before my eyes. He had been swallowed by the bottom! While I was trying to figure out if I was hallucinating from diver's narcosis, Manu's legs disappeared. Bubbles rose up from the bottom; they were Kike's. I came closer and realized that the bottom was not the bottom, but a layer of colder, murky water, hyper-loaded with sediment. A ghostly Kike appeared from the cloud, and I laughed so hard that I almost lost my regulator.

Back at the surface we saw a skiff with two French spearfishermen carrying long wooden spear-guns. They told us that they target all these rocks within a one-day distance from Libreville, and that

even trawlers come here at night and drag their nets over them. And even though we didn't ask, they told us they don't shoot the large groupers. In my opinion, though, it was too evident that they do. No wonder we had not seen any near Libreville.

We spent two weeks scuba diving and exploring deep reefs with the ROV, discovering forests of soft corals, meadows of free-living calcareous algae, and rare giant groupers. And we realized that fishing, legal but mostly illegal, was threatening Gabon's marine resources. But the most memorable moment occurred the last day of the expedition.

An Influential ROV Driver

The vessel was anchored on top of a seamount 91.5 meters deep, 40 kilometers offshore. The sea was flat as a mirror, the sun so bright that we could not be on deck without sunglasses. At noon, Mike Fay and Lee White, director of Parcs Gabon, the country's national park agency, arrived on board, accompanied by President Ali Bongo Ondimba and

HOME SWEET HOME
A moray eel peers out from its cup-coral home, a reef growing around the beams of an oil platform off the coast of Gabon. Such platforms are artificial habitats that support reefs that would otherwise be absent from the sandy Gabonese coast. *(page 166)*

THE MOST PRISTINE BEACHES IN AFRICA
Remote and undeveloped, Gabon's beaches harbor extraordinary wildlife. Forest elephants and hippos are regular visitors at Loango National Park's beaches.

his wife, Sylvia. We took them to the ROV room, so full of large monitors it looked like NASA's Mission Control center. Mike introduced the Pristine Seas team, and I showed the highlights of our expedition to the president. After a lively discussion of about an hour, the president looked at his watch. I feared we were going to lose his attention—he was supposed to chair a cabinet meeting later that day. And then Joe Lepore, dive safety officer of the Waitt Foundation, asked the president if he'd like to drive the ROV. The president's eyes lit up. He did pretty well for a beginner ROV operator—definitely better than I had! After a few minutes of flying over a sandy flat bottom, the ROV camera revealed a dark area in the background. The president piloted the ROV closer, and the shadows turned into a deep reef, made of dark volcanic rocks full of little caves and overhangs. Over the rocks swam hundreds of pink basslets, while red scorpionfish lay motionless, surrounded by thousands of brittle stars. It was by far the most gorgeous reef we had seen during the entire expedition. As we were rejoicing over the view, a dogtooth grouper the size of a car's windshield appeared on the right side of the screen. Another of these gray giants appeared from the right. Moving slowly, as if they were timidly curious, three more groupers came into the frame. The president, now wearing a red National Geographic cap, had a smile so large it did not fit in his face. Seeing firsthand what few Gabonese have seen only reassured him that something had to be done to protect these underwater treasures.

Finally, the president did leave the vessel, and afterward we brought the ROV back to the surface. On the way up, a couple of silky sharks approached the ROV, one with a hook and line on its mouth, and a school of bigeye jacks swam in circles around it. We spent one more hour exploring with the ROV before we decided it was time to sail back to Libreville. That was the end of the expedition, and we had reason to believe that we had impressed someone enough to make a difference. ▪

COASTAL GABON

SÃO TOMÉ & PRÍNCIPE

PRÍNCIPE

São Tomé
SÃO TOMÉ

SAO TOME & PRINCIPE

Annobón
Equatorial Guinea

EQUATOR

2°S

4°S

6°E 8°E 10°E 12°E

EQUATORIAL GUINEA
RÍO MUNI

Oyem

Makokou

Cabo San Juan Calatrava
Islas Elobey
Corisco Bay
Corisco
Boundary claimed by Gabon
Boundary claimed by Equatorial Guinea

Baie de
Mondah
AKANDA
NATIONAL PARK

Pointe Pongara Libreville

PONGARA
NATIONAL PARK

WONGA-WONGUÉ
PRESIDENTIAL
RESERVE

Bifoum

G A B O N

Ogooué

Cap Lopez
Port-Gentil

Lambaréné

Koulamoutou

Golfe
d'Olindé

Ogooué

Moanda

Lagune
Nkomi
Omboué

LOANGO
NATIONAL
PARK

Mouila

MOUKLABA-DOUDOU
NATIONAL PARK

Iguéla
Lagune
Iguéla

PETIT-LOANGO
FAUNAL RESERVE

NGOVE-
NDOGO
HUNTING
AREA

Ndendé

Setté Cama
Lagune
Ndogo

Gamba

SETTÉ-CAMA
HUNTING AREA

Tchibanga

Mossendjo

OUANGA PLAIN
FAUNAL RESERVE

C O N G O

Mayumba

MAYUMBA
NATIONAL
PARK

Dolisie
(Loubomo)

Pointe-
Noire

Tshela

Cabinda
Angola

Cabinda

DEMOCRATIC
REPUBLIC
OF THE
CONGO

Boma

Muanda

Mouth of the Congo River

Soyo

A N G O L A

MAP KEY

- - - - - Maritime treaty boundary

——— Selected park on Gabonese coast

Depth in meters
Sea Level

100
250
500
1000
1500
2000
2500
3000
3500
4000
4500
5475

GABONESE EXCLUSIVE ECONOMIC ZONE

Sao Tome & Principe
Equatorial Guinea

Sao Tome & Principe
Gabon

Ten marine protected areas covering
47,000 square kilometers were announced
by President Ali Bongo Ondimba in November 2014.
The boundaries of these protected areas had not
been finalized at the time of the printing of this book.

Mercator Projection

SCALE 1:3,800,000

1 CENTIMETER = 38.0 KILOMETERS; 1 INCH = 60.0 STATUTE MILES

0 25 50 75 100
KILOMETERS

0 25 50 75 100
STATUTE MILES

0 25 50 75 100
NAUTICAL MILES

Europe

Africa

Coastal
Gabon

South
America

ATLANTIC
OCEAN

EQUATOR

THE GABONESE REPUBLIC, a country on the west coast of central Africa, is famous for its 13 national parks that protect pristine tropical forests. The coast of Gabon is more than 800 kilometers long, composed mostly of sandy beaches, mangroves, and coastal lagoons. The pristine, uninhabited beaches of Gabon offer the rare sight of hippos surfing near shore and humpback whales breaching in the distance. Offshore, dozens of oil platforms act as artificial reefs, attracting fish and other reef animals. The Ogooué and the Congo Rivers discharge an enormous amount of sediment and nutrients, making Gabon's coastal ocean very productive.

RAINBOWS GLISTEN
A school of rainbow runners aggregates around an oil platform off the coast of Gabon. These platforms act as fish aggregators, and more than one ton of fish will congregate in the first 20 meters of depth below.

LIFE IN EVERY CORNER
A combtooth blenny peeks out from its tiny hiding place, having moved into the empty shell of a barnacle that once grew and attached to an underwater beam of a Gabonese oil platform. *(following pages)*

"I'm grateful for National Geographic Pristine Seas' CGI [Clinton Global Initiative] commitment to double the size of protected ocean areas. They're well on their way, protecting areas of exceptional biodiversity, while rebuilding fish populations and boosting tourism. An extraordinary achievement."

—PRESIDENT BILL CLINTON

GHOSTS OF THE SHIPWRECK
The Pristine Seas divers found schools of drums living in shipwrecks off Libreville, the capital of Gabon. Surrounded by muddy bottoms and murky waters, shipwrecks are artificial reefs full of marine life.

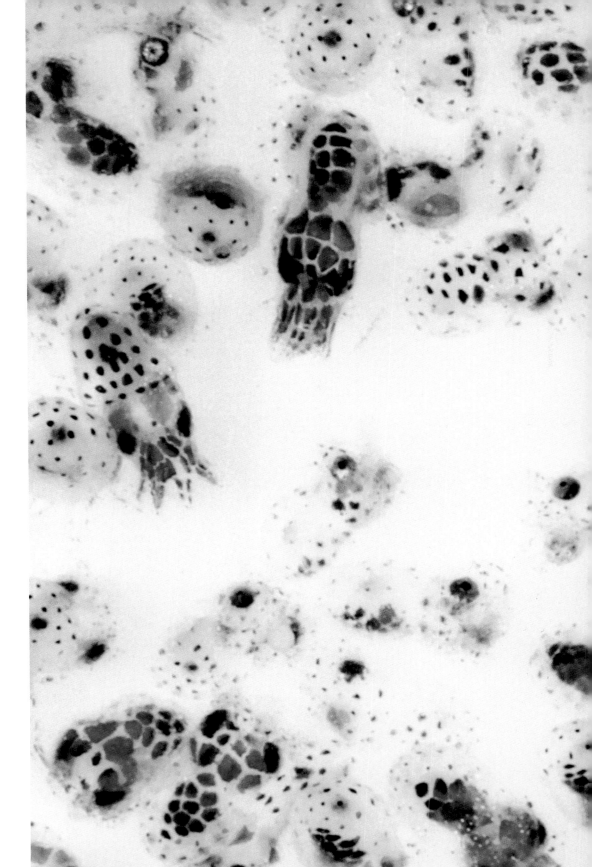

A MARINE MATERNITY WARD
Using the Waitt Foundation's remotely operated vehicle, the Pristine Seas team pulled an empty shell from under the Gabon water and placed it in a tank on board. Minutes later, thousands of baby octopuses hatched from eggs that had been laid inside the shell.

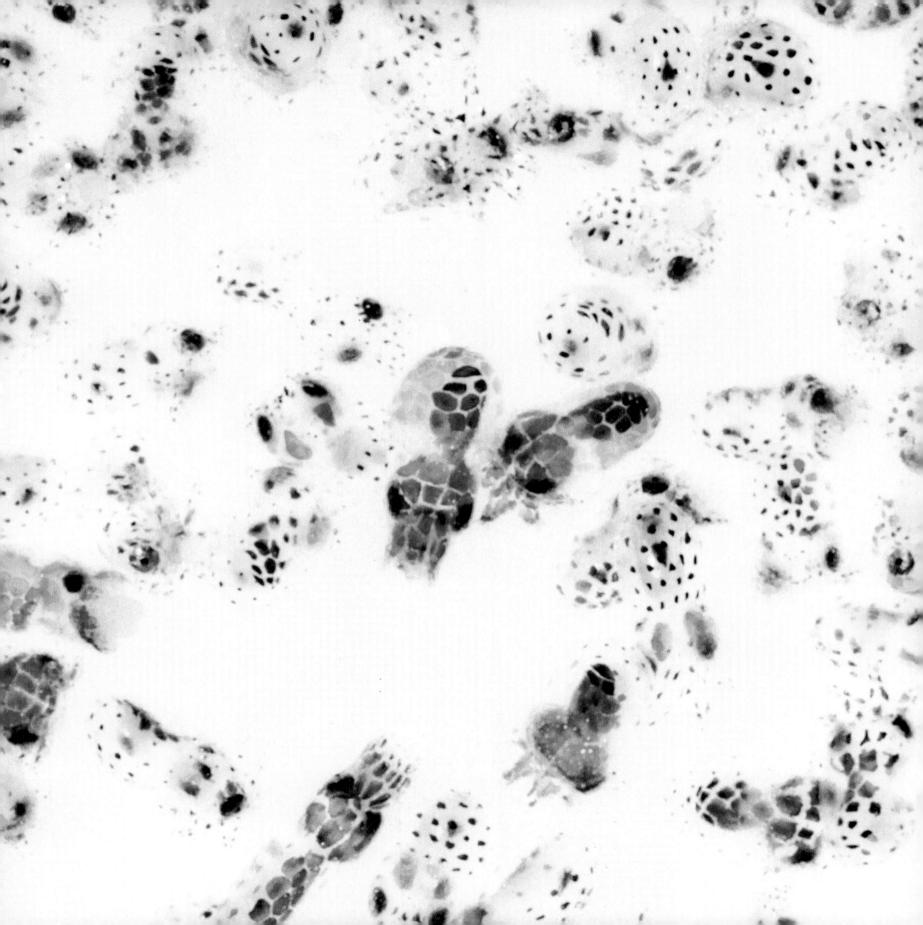

OASIS IN AN UNDERWATER DESERT
Most of Gabon's coast is sandy or muddy,
but here and there rocky reefs and ledges
provide the perfect habitat for gorgonians
and sponges to thrive.

A TREASURE ON AN INDUSTRIAL PLATFORM
Offshore oil platforms in Gabon contain more
marine species per square meter than the soft
bottoms surrounding them. Here we can see
a spiny lobster, creolefish, sponges, barnacles,
bryozoans, oysters, cup corals, and a pencil
sea urchin.

KEEPING A CLOSE EYE ON VISITORS
Silky sharks came to inspect the Waitt Foundation's remotely operated vehicle during the 2012 Pristine Seas expedition to Gabon. We took the presence of sharks to be a good sign in this region, known for its dismal fisheries management and an absence of conservation measures.

Unlikely Havens

Offshore oil is the most valuable economic resource for Gabon, yet the president was committed to protect Gabon's waters. Was it possible to do one without jeopardizing the other? To resolve this paradox, we needed to see what oil platforms did to the marine environment. We had to go diving in hell—or so we were told.

One morning in October 2012 we approached our first oil platform, anchored in water more than 42 meters deep off the central coast of Gabon. The platform's footprint was that of a large suburban home. The operating quarters, 12 meters above the surface, were a three-dimensional maze of pipes and valves of all types and sizes, populated by a handful of workers. Metallic bangs and other noises provided an unpleasant backdrop. A long black pipe projected away from the platform like a dragon's neck and exhaled fire. The oil company's safety officers opposed our diving there, but because we obtained permission from the higher-ups in the company, they had to oblige us. Nevertheless, they warned us that it was extremely dangerous to dive under the platform, that underwater pipes would suck us in, and that the flare would burn us. A litany of horrors.

As soon as we dived in, all our worries disappeared. Immediately, the magic captivated us. Because it was 32 kilometers offshore, this water was clear, unlike the muddy coastal waters. A flotilla of jellyfish was stationed around the platform, each animal with its own school of small fish moving nervously among its tentacles like a silver rain. A school of barracuda was hovering next to the large pillars supporting the platform. On the other side of the platform we saw a school of rainbow runners and jacks. The beams supporting the structure were fully covered by orange cup corals, algae, barnacles, and oysters. There was not a single square centimeter that wasn't occupied by a plant or an animal, including spiny and slipper lobsters, millions of brittle stars, and large patches containing the purple eggs of damselfish. A 1.5-meter-long African brown snapper shot up from the deep toward Kike Ballesteros, but when it realized that Kike was slightly bigger than it was, the snapper bolted back into the deep. During the entire dive we heard the ethereal singing of humpback whales. Back on the surface, our boat driver told us that two humpback whales had been breaching a few hundred meters from the platform. Diving in hell wasn't so bad after all. ■

WHERE OFFSHORE OIL MEANS LIFE
As shown by this underwater view of an offshore oil platform in Gabon,
every square centimeter of the structure is covered with cup corals or other
bottom-dwelling animals necessary to support a healthy fish assemblage.

Until the early 1990s sea ice was found throughout the Franz Josef Land archipelago. But when we were there in the summer of 2013, there was no sea ice to be found.

ARCTIC WATERS

Franz Josef Land

Russia's Far North

My fingers were freezing. I was in so much pain that I couldn't feel if I was still holding my camera. My lips were so numb that I could barely hold the regulator in my mouth. The water was green and murky, and I could not even see my fins. I continued descending until the green water disappeared, as though I were flying through a cloud and reaching the other side, where the sky opens up. A secret universe unveiled before my eyes. A forest of brown algae was growing on an underwater slope of dark volcanic rock. Below the algae, a bed of pale anemones looked like a flower meadow. I swam closer to the rock. An orange sea spider the size of my hand moved slowly through the undersea meadow, in search of prey. After five years of dreaming and two years of preparation, I was diving at Franz Josef Land, virtually the top of the world.

Franz Josef Land is the northernmost archipelago in Russia and the closest landmass to the North Pole. Only the northern tip of Greenland reaches farther north. In the winter, ice and snow carpet the 192 islands of Franz Josef Land. The sea freezes, locking the islands in a white trap. The sun is gone for six months, submerging this remote world in perpetual night. It is one of the most forbidding places on the planet. In the summer, when the sea ice melts and the sun stays above the horizon 24 hours a day, the cold and the darkness morph into a blooming miracle, turning white ice into a green oasis.

After 45 minutes underwater, I started to shiver. It was time to return to the surface. As I surfaced, the regulator came out of my mouth, and my lips felt huge. I asked our boat driver to help me with my camera, and he looked at me, frowning. I wasn't making much sense. It was already difficult for a Catalan-born man speaking Spanglish to communicate with a man speaking only Russian, and my numb lips did not help. I raised my camera, and he understood. He lifted the camera into the boat, while I tried to unclip my diving buoyancy compensator device and my diving tank. I just couldn't manipulate anything with my frozen, painful hands in my thick neoprene gloves. He saw my pathetic attempts and reached out, alleviating my comical misery.

Signs of Change

I climbed back into the Zodiac dinghy, exhaled deeply, and looked up. Above us was Rubini Rock, a massive vertical cliff made by basalt that had crystallized in the shape of hexagonal prisms, giving it the look of a giant pipe organ. The squeaks and calls of the seabirds were deafening. From a distance, the thousands of birds flying off the cliff seemed like a giant cloud of mosquitoes on a summer day.

Up close, kittiwakes nested on the rocky cliff and slope, taking advantage of every crack and nook, every hole beneath boulders, to lay their eggs and raise their chicks. Little auks nested nearby as well, looking like cute penguins that can fly. Every single horizontal space, many as small as a dessert plate, was occupied. The opening of the sea ice in the summer and the permanent sunlight fosters blooms of microscopic organisms in the sea, which are the base of the food chain that supports such an extraordinary number of seabirds.

But now, because of global warming, Arctic waters are getting warmer, which facilitates the northward migration of plankton from the Atlantic. Little shrimplike organisms, copepods, are the main prey item for the little auks. An Atlantic species of copepod, smaller and less nutritious than its Arctic counterparts, is now found in increasing numbers around Franz Josef Land, and the result may have serious consequences for the birds nesting here. If Atlantic copepods replace Arctic copepods, the little auks' feeding habits, and hence their ability to raise chicks, may be compromised. This is just one of many changes occurring in Franz Josef Land and the high Arctic. We do not know what the outcome will be, but we do know that the Arctic of the future will be different from that witnessed by the 19th-century explorers who discovered these islands.

Until the early 1990s sea ice was found throughout the Franz Josef Land archipelago. But when we were there in the summer of 2013, there was no sea ice to be found. We were standing at Cape Fligely on Ostrov Rudolph (Rudolf Island), the northernmost point in Franz Josef Land, and all we could see was blue water as far as the horizon. Near our ship we saw a polar bear standing on black rock between the edge of a glacier and the sea. We put one of our small boats in the water and motored closer to the bear. Behind the massive animal, the glacier was melting, sending a waterfall arcing into the sea. It seemed as if the bear was also looking for the sea ice, confused by the changes in its world. The floe edge—in essence, the glacial coastline where ice meets the sea—was nowhere to be seen. Because of that the seals are gone, and the polar bears, which traditionally feed on seals, have to depend on a diet based mostly on seabirds and grass. As with the little auks, the changing climate is reshaping the polar bears' diets and compromising their ability to raise young.

To Dive With a Walrus

I love diving with sharks and other large animals. Diving with whales is especially exciting because they are the most massive animals in the ocean. Being close to them brings me closer to life—it is hair-raising yet safe. But there was one large animal at Franz Josef Land that I wasn't so sure about diving with: the walrus.

Despite the lack of sea ice, Franz Josef Land continues to be an extraordinary refuge for the Atlantic walrus. To put it bluntly, these animals are mountains of fat guarded by long ivory tusks. Large males weigh up to 1,360 kilograms. I can imagine the first humans who saw walruses; they might have thought they were marine monsters. We came to Franz Josef Land with the intention of diving with them.

Near the old Soviet Tikhaya research station, we donned our dry suits and carried our tanks and cameras down a precarious ladder from our mother ship to our Zodiac. We then motored west for half an hour, toward a small island where we were told we would find a walrus aggregation.

We could smell them before we could even see them. About 300 walruses were there, lying down lazily on the beach—males, females, and babies we estimated to be two months old. We approached in the Zodiac, as quietly as possible. A few walruses that were already in the water came out to inspect us. Walrus whiskers seem like a thick monofilament fishing line, but they are very sensitive organs. We cautiously put our cameras in the water with a pole, to get a feel for their behavior before jumping in. A group of four walruses came over immediately to inspect our boat. As they got closer, we realized how large they were. An immense female swam close to us and touched the camera dome with her whiskers. She swam back suddenly, surprised and apparently scared by the feeling of the glass on her whiskers. The other walruses dived and disappeared.

We waited for them to surface. I became a little anxious as we saw two of them diving underneath our Zodiac. Suddenly, a male surfaced so close to us that its exhale sprayed me. Lifting a third of its body out of the water, he looked like a mountain with tusks. He dived again and swam away. At that very moment we decided that we were going to postpone our dive with the walruses.

Arctic Accomplishments

We spent four weeks sailing a total of 3,500 kilometers and surveying 24 of the Franz Josef Land islands. We landed at 35 different spots. We observed 45 whales, 42 polar bears, and more than 3,000 walruses, mostly in large colonies on beaches. We deliberately photographed the identical scenery found in 180 historical photographs, taken in the early 20th century. Our intention was to document the changes that had occurred in the environment of Franz Josef Land since the first explorers arrived here in the late 1800s.

With this expedition, we expanded scientific knowledge of a pristine sea rarely visited by humans, let alone divers. We dramatically increased the span of underwater exploration of Franz Josef

AN ANGEL IN THE ARCTIC
Clione limacina, the sea angel, is an Arctic sea slug that is common in the surface waters of Franz Josef Land. When we jumped into the freezing water, the sea angels made us feel as if we were in heaven. *(page 190)*

AN EXPANSE OF MELTING ICE
When polar satellite observations started in the 1970s, much of Franz Josef Land's waters were frozen year-round. Now the sea ice melts away every summer.

Land. Where before only four main spots had been occasionally dived at, we conducted 229 dives at 22 diving sites, spending a total of 111 hours underwater. And we also broke new ground by exploring the deep waters, down to a depth of 392 meters, with our National Geographic "drop-cams." The drop-cams are hollow glass balls the size of a giant watermelon with a high-resolution video camera and lights inside. It's our cheap alternative to a submersible. We dropped them at 24 spots, and made an exciting discovery: a Greenland shark, the first ever seen at Franz Josef Land.

Franz Josef Land remains an inhospitable environment in the winter, but the warmer summers and the loss of sea ice are making it more appealing to visitors. Human tourists will come seasonally, but other species coming from the south, such as warmer-water plankton and Atlantic cod, might stay, bringing with them unpredictable changes to a place that had been isolated for thousands of years. But we can help Arctic wildlife in Franz Josef Land

to buffer the impacts of global warming by preventing direct human threats—mining and drilling, fishing, pollution—and thus preserve this jewel of the Russian Arctic.

Franz Josef Land is a *zakaznik,* a federal reserve managed by the Russian Arctic National Park. After the expedition, the park submitted a proposal to the Russian Ministry of Natural Resources and the Environment to give Franz Josef Land and the waters surrounding it national park status. In a time when the loss of summer sea ice in the summer will also see an increase of shipping, oil exploration, and development, Russia has a unique opportunity to fully protect one of the last places left on the planet that still resembles the world during the last ice age. Franz Josef Land is a living experiment, a place to observe how Arctic wildlife will adapt to a warming world. Ensuring that this precious refuge remains safe from direct human impacts will show Russia's leadership in Arctic conservation and send a strong signal to other nations to follow. ▤

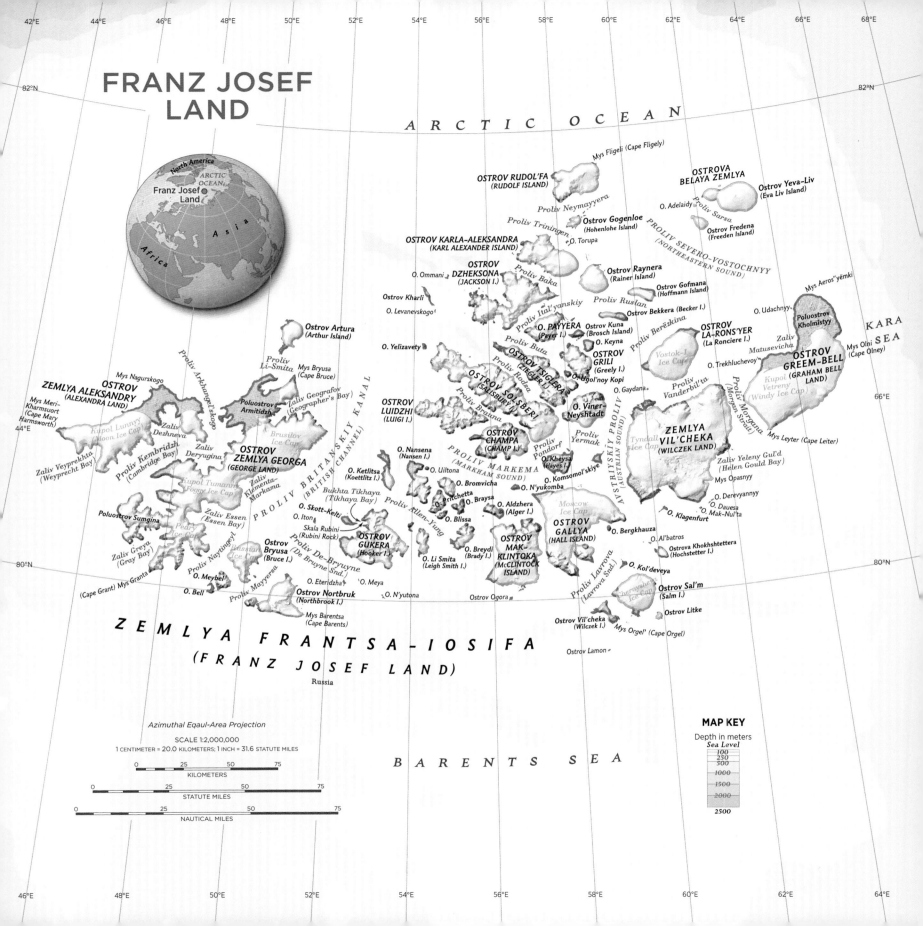

FRANZ JOSEF LAND

ARCTIC OCEAN

Mys Fligeli (Cape Fligely)

OSTROV RUDOL'FA
(RUDOLF ISLAND)

**OSTROVA
BELAYA ZEMLYA**

Proliv Neymayyera

O. Adelaidy   Proliv Sarsa   **Ostrov Yeva-Liv**
(Eva Liv Island)

Proliv Triningen   **Ostrov Gogenloe**
(Hohenlohe Island)   Ostrov Fredena
(Freeden Island)

OSTROV KARLA-ALEKSANDRA
(KARL ALEXANDER ISLAND)   O. Torupa

Proliv Baka   PROLIV SEVERO-VOSTOCHNYY
(NORTHEASTERN SOUND)

**OSTROV
DZHEKSONA**
(JACKSON I.)   O. Ommani   **Ostrov Raynera**
(Rainer Island)

Ostrov Kharli   Ostrov Gofmana
(Hoffmann Island)

Mys Aeros"yëmki

O. Levanevskogo   Proliv Ruslan   Ostrov Bekkera (Becker I.)   O. Udachnyy

Poluostrov
Kholmistyy   **KARA
SEA**

Ostrov Artura
(Arthur Island)   O. Yelizavety   Proliv Ital'yanskiy   **O. PAYYERA**
(Payer I.)   Ostrov Kuna
(Brosch Island)   Proliv Berëzkina   **OSTROV
LA-RONS'YER**
(La Ronciere I.)   Zaliv
Matusevicha   **OSTROV
GREEM-BELL**
(GRAHAM BELL
LAND)

Proliv
Li-Smita   Proliv Buta   O. Keyna   O. Trekhluchevoy   Mys Olni SEA
(Cape Olney)

Mys Nagurskogo   Mys Bryusa
(Cape Bruce)   **OSTROV
GRILI**
(Greely I.)

ZEMLYA ALEKSANDRY
(ALEXANDRA LAND)   **OSTROV
SOL'SBERI**
(SALISBURY I.)   Vostok-1
Ice Cap   Proliv
Vanderbil'ta   Kupol
Vetreny
(Windy Ice Cap)

OSTROV TSIGLERA
(ZIEGLER I.)   O. Ugol'noy Kopi   O. Gaydana

Poluostrov
Armitidzh   Proliv Rodsa

Mys Meri-
Kharmsuort
(Cape Mary
Harmsworth)   Zaliv Geografov
(Geographer's Bay)   Proliv Brauna   **O. Viner-
Neyshtadt**   Proliv Morgana
(Morgan Strait)   Mys Leyter (Cape Leiter)

Kupol Lunnyy
(Moon Ice Cap)   Zaliv
Dezhneva   Brusilov
Ice Cap   **OSTROV
LUIDZHI**
(LUIGI I.)   **OSTROV
CHAMPA**
(CHAMP I.)   Proliv
Yermak   **ZEMLYA
VIL'CHEKA**
(WILCZEK LAND)   Zaliv Yeleny Gul'd
(Helen Gould Bay)

Zaliv
Deryugina   Proliv
Pondorf   Mys Opasnyy

Zaliv Veyprekhta
(Weyprecht Bay)   Proliv Kembridzh
(Cambridge Bay)   Proliv Markema
(MARKHAM SOUND)   Tyndall
Ice Cap

**OSTROV
ZEMLYA GEORGA**
(GEORGE LAND)   Zaliv
Klementa-
Markama   O. Nansena
(Nansen I.)   O. Kheysa
(Hayes I.)   O. Derevyannyy

Kupol Tumannyy
(Foggy Ice Cap)   O. Ketlitsa
(Koettlitz I.)   O. Uiltona   O. Komsomol'skiye   O. Dauesa
O. Mak-Nul'ta

Bukhta Tikhaya
(Tikhaya Bay)   O. Bromvicha   O. N'yukomba

Poluostrov Sumgina   O. Pritchetta   Moscow
Ice Cap   O. Klagenfurt

Peary
Ice Cap   Skott-Kelti   Proliv Allen-Yung   O. Braysa   **OSTROV
GALLYA**
(HALL ISLAND)   O. Bergkhauza   O. Al'batros

O. Iton   O. Aldzhera
(Alger I.)   Ostrova Khokhshtettera
(Hochstetter I.)

Zaliv Essen
(Essen Bay)   Skala Rubini
(Rubini Rock)   O. Blissa   **OSTROV
MAK-
KLINTOKA**
(McCLINTOCK
ISLAND)

Zaliv Greya
(Gray Bay)   Russkiy
Ice Cap   **OSTROV
GUKERA**
(Hooker I.)   O. Breydi
(Brady I.)

**Ostrov
Bryusa**
(Bruce I.)   Proliv De-Bryuyne
(De Bruyne Snd.)   O. Li Smita
(Leigh Smith I.)

O. Meybel   O. Meya   Proliv Lavrova
(Lavrova Snd.)   Cherayshen
Ice Cap   **Ostrov Sal'm**
(Salm I.)

(Cape Grant) Mys Granta   O. Eteridzha   O. Kol'deveya

O. Bell   O. N'yutona   **Ostrov Nortbruk**
(Northbrook I.)   Ostrov Ogora   Ostrov Litke

Proliv Mayyersa   Mys Barentsa
(Cape Barents)   **Ostrov Vil'cheka**
(Wilczek I.)   Mys Orgel' (Cape Orgel)

ZEMLYA FRANTSA-IOSIFA
(FRANZ JOSEF LAND)

Russia   Ostrov Lamon

Azimuthal Eqaul-Area Projection

SCALE 1:2,000,000

1 CENTIMETER = 20.0 KILOMETERS; 1 INCH = 31.6 STATUTE MILES

0   25   50   75
KILOMETERS

0   25   50   75
STATUTE MILES

0   25   50   75
NAUTICAL MILES

BARENTS SEA

MAP KEY

Depth in meters
Sea Level
100
250
500
1000
1500
2000
2500

FRANZ JOSEF LAND is the northernmost archipelago in Eurasia. Located between latitudes 80° and 82° N, it is composed of 192 islands with a land surface of 16,134 square kilometers. Glaciers cover 85 percent of the land. The waters surrounding the archipelago are covered by sea ice half of the year, and even summer seawater temperature here can be as low as minus 2°C. Abundant marine mammals—bowhead whales, narwhals, seals, walruses, and polar bears—and hundreds of thousands of seabirds live here. There are 43 species of fish, none of which is abundant or currently exploited commercially.

WALRUSES GATHER IN ICY WATERS
During a Franz Josef Land summer, walruses aggregate in rookeries: one-ton males, females, and newborns. They feed on clams that live buried in the muddy bottoms close to shore.

THE FROZEN GARDEN
Franz Josef Land's rocky reefs are covered by a thick kelp canopy, below which anemones grow like underwater flowers. Mysid shrimp swarm like bees over the ocean bottom.

IN SEARCH OF SEA ICE
A solitary polar bear watches the ice-free sea in August 2013, backed by a glacier at Cape Fligely, the northernmost land in Eurasia. *(following pages)*

"For the present we must stay where we were and anoint ourselves with the ointment called Patience, a medicament of which every polar expedition ought to lay in a large supply."

—FRIDTJOF NANSEN

CORAL COME ALIVE?
The branching shape of a nudibranch gives the impression of an animated piece of coral. More than 3,000 known species have been identified, coming in all shapes, sizes, and colors and found all over the world—even in the icy waters of Franz Josef Land.

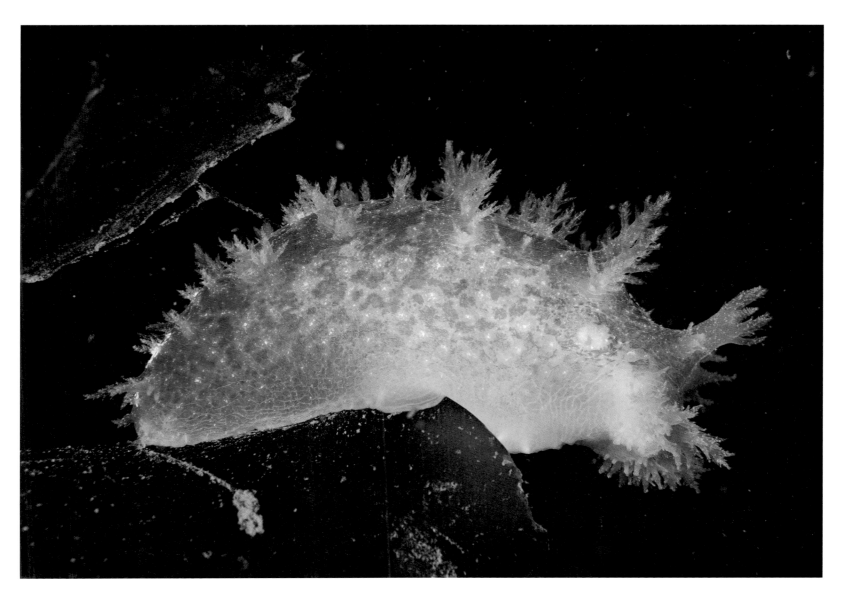

SLOW PACE ALONG THE MUDDY BOTTOM
Nudibranchs are carnivores without picky tastes. They can
even eat toxic sponges or anemone stingers and then store
the poison in their own bodies to help repel predators.

A SEA FULL OF STARS
Brittle stars are the most common animal on muddy bottoms off Franz Josef Land. They are versatile consumers, scavenging dead animals or raising their tentacles to filter feed.

Sea Ice
No More

In 1893 the Norwegian explorer Fridtjof Nansen attempted to reach the yet inaccessible North Pole by doing something that other polar explorers thought crazy, even suicidal. He had a ship, the *Fram,* built to withstand the pressure of the sea ice, with a rounded hull and keel that would, in Nansen's words, "slip like an eel out of the embraces of the ice" instead of being crushed like many other ships before it. Nansen and his legendary crew sailed the *Fram* to northern Siberia and let the sea ice freeze around them. His theory was that the polar currents would carry the ship to the North Pole. The *Fram* drifted west but never got close to the geographical Pole. Impatient, Nansen and his companion, Hjalmar Johansen, took off with sledges and a team of dogs to reach the North Pole on foot. Because of intractable ice, they turned south at latitude 86° 13.6′ N and reached Franz Josef Land in August. In September 1895 they decided to overwinter on what is now known as Cape Norwegia. For almost eight months they lived in a small hut dug a meter into the ground and covered with walrus hides. They slept in a single sleeping bag and ate polar bear meat cooked in walrus blubber.

Their story caught my imagination as no other before. Once I started reading Nansen's account, *Farthest North,* I could not stop thinking about our own expedition to Franz Josef Land. I decided to obtain all available summer photographs taken by Nansen and other early explorers. Kristin Rechberger photographed the same scenes during our 2013 expedition. Wonderful archivists at Oslo's National Library of Norway kindly allowed me to inspect Nansen's materials, including handwritten notes and his field journals stained with walrus fat. We made high-resolution copies of Nansen's and other photographs, including those from Norwegian historian Susan Barr, who made rare visits to Franz Josef Land in the early 1990s.

We came home with a long-term gallery of change. The most striking change of all: Nansen skied on sea ice and snow between islands in August 1895. Even in August 1995 Barr photographed abundant sea-ice floes. But in August 2013 nowhere in Franz Josef Land—not even at Rudolf, the northernmost island in the archipelago—did we see any sea ice. The loss of sea ice is one of the most dramatic changes on our planet occurring as a result of global warming, and we are only starting to understand the consequences for Arctic wildlife—and ourselves. ∎

CHANGE OVER TIME
When the Ziegler Polar Expedition visited Franz Josef Land in August 1903, they found a frozen channel between McClintock Island and Alger Island. In 2013, at that same place, we could sail through open water. The only ice to be seen came from icebergs that had broken off glaciers.

A FISH IN COLD WATER
A juvenile Atlantic spiny lumpsucker sits on a kelp frond off Franz Josef Land. In these extremely cold waters, fish are very rare.

EXPLORING THE NORTHERN FOREST
Underwater camera assistant Nathan Lefevre dives in the northernmost kelp forest in the world, off Cape Fligely on Rudolf Island. The loss of sea ice in the summer might benefit the growth of kelp forests in these high Arctic latitudes. *(following pages)*

"The day will come when men like me will see the murder of an animal as they now see the murder of a man."

—LEONARDO DA VINCI

HIGH-DENSITY HOUSING
Every nook and cranny of the basaltic cliff of Rubini Rock is occupied by a nesting pair of seabirds, mostly kittiwakes. Summer in Franz Josef Land brings nutrients and light to surface waters, which bloom with plankton that supports fish and seabirds alike.

PERMANENT ATTACHMENT

A stalked jellyfish just a few centimeters long lives attached to a kelp blade for the duration of the Arctic summer. Other jellyfish species float free, but the stalked jellyfish must attach to hard substrates.

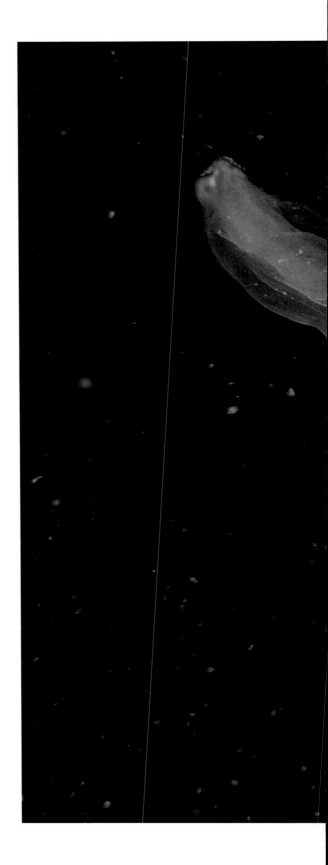

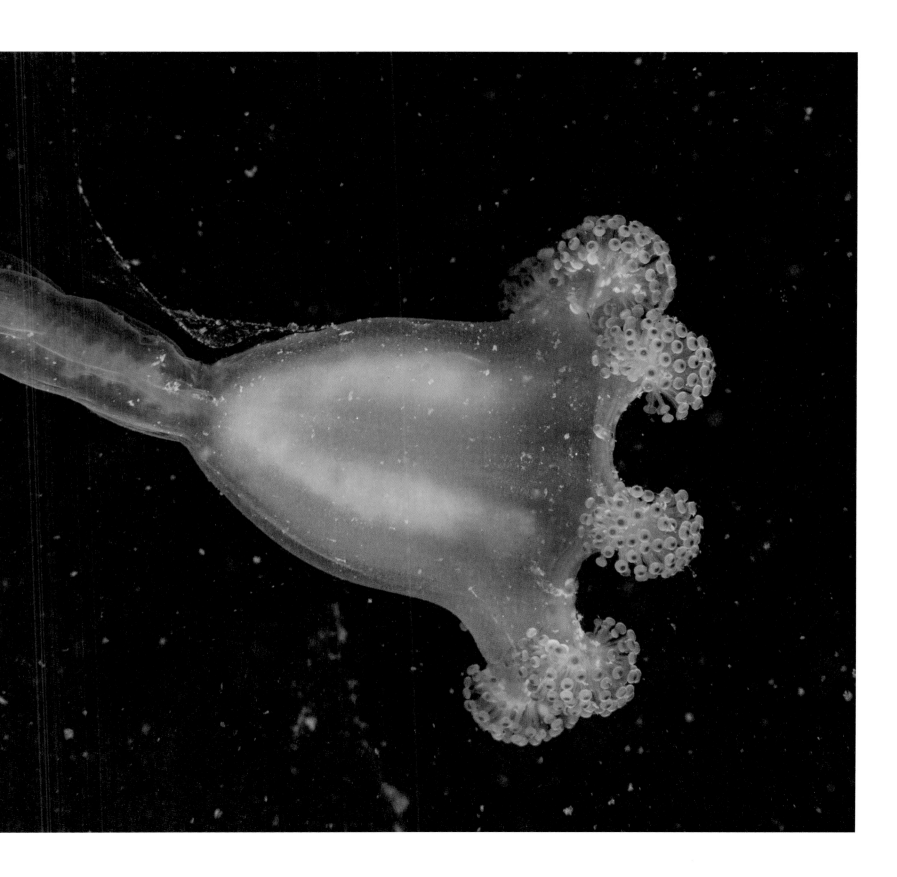

BELOW THE TIP OF THE ICEBERG
Just minutes after this Pristine Seas diver
returned to the surface, a larger iceberg broke
and capsized—a catastrophe luckily averted.

Veggie
Polar Bears

August 2013. We were standing at Cape Fligely on Ostrov Rudolph (Rudolf Island), the northernmost point in Franz Josef Land and the closest land to the North Pole in Eurasia. Russian polar explorer Victor Boyarsky had told me that Cape Fligely was his favorite place in Franz Josef Land. "Standing on the cliff and looking north you get the feeling of Arctic wilderness, with ice covering the sea as far as the North Pole," he had said in March 2013. But now, in the summer, all we could see was blue water as far as the horizon. This is happening throughout the Arctic—the most conspicuous impact here of climate change.

The week before, the slopes of the small Ostrov Brosch (Brosch Island) were covered with green. From the deck of our ship I saw a white dot moving slowly on the cliff. We grabbed our cameras, jumped into a Zodiac, and zoomed to the island. As we got closer, we realized that the white dot was a polar bear lifting and moving boulders around. All of a sudden the bear dived into a hole in between boulders. A little auk flew out of the hole like a rocket, avoiding the bear's terrible jaws. The bear turned around, its paws stained with bird blood. It felt to us as if the bear itself understood how sad and pathetic a scene it was, to see a polar bear, the great carnivorous predator of the Arctic, failing to capture one little bird to eat.

Seals—the polar bear's main prey—eat fish that eat little shrimp that eat microscopic algae growing below the ice, near the edge of the ice pack. With the sea-ice edge nowhere to be seen in summer 2013, a cascading effect was changing the habitat's food chain. Without the algae, the shrimp were depleted. Without the shrimp, the seals were gone. Without the seals, polar bears had to search for other food for their survival. Walruses were still present in Franz Josef Land, but they are too massive and dangerous, and thus polar bears approach them with caution. All that was left was grass and seabirds, resulting in an almost vegetarian alternative for the largest carnivore in the Arctic. ■

DISCOVERING GREEN GRASS
A playful polar bear relaxes on the grassy slopes of Brosch Island
after eating a meal of little auks, small seabirds common to the area.

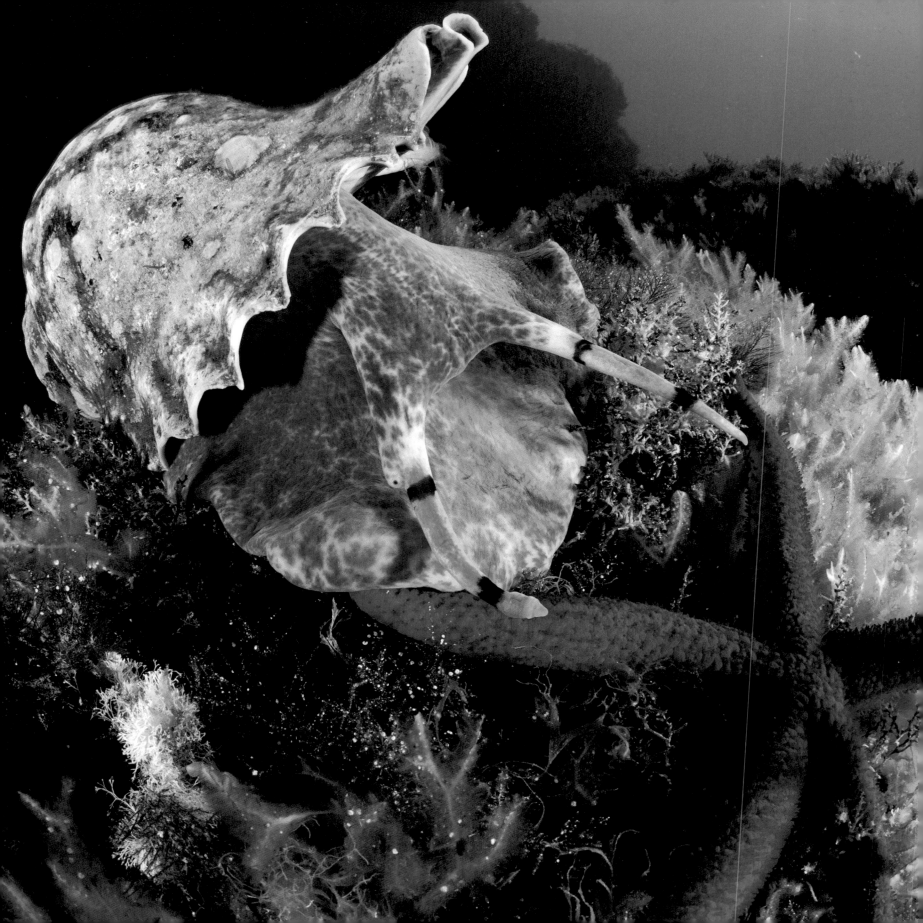

Reserves are not the only tool to restore the health and productivity of the ocean, but we know that, if well managed, they work extraordinarily well, for marine life and for people.

TOWARD THE FUTURE

The Mediterranean and Cuba

Beacons of Hope

Television got me into the outdoors. That hypnotic box had me dying to go out and explore. But before exploring, I had to dream—and Jacques Cousteau's TV shows gave me what I needed to feed my curious mind. Thanks to them, I could begin to imagine all the travels and discoveries I might be able to carry out when I grew up. I went back and forth, from the diving adventures taking place in my living room to my own adventures, swimming in the coves of the Mediterranean Sea near my home. It was a schizophrenic journey, for Cousteau and his fellow divers swam among sharks and large fish, whereas my backyard sea was empty.

When I was a child, I didn't wonder why the Mediterranean was so empty of fish and other underwater life. I thought that a poor Mediterranean was natural. I did not know it, but our sea was dying. Only when my television hero, Cousteau, started to warn us about the effects of too much fishing and pollution on the oceans did I realize my ignorance. A story was building up in my young mind, and as it grew I began to see it replicated everywhere I looked. Where once there had been riches, now there lay a barren sea. We were pulling fish out of the water faster than they could reproduce, and we were throwing back into the water what we didn't want: trash and sewage.

The word "pristine" means "in its original condition" or "unspoiled." Pristine areas in the ocean are pristine because, by definition, there are no humans living around them, or so few they make no visible impact. Pristine places are nearly as rare as unicorns, and we need to protect them, and through the Pristine Seas project, I believe we are taking important steps to do so.

But what about the rest of the ocean, where people build, fish, and pollute? The question looming in my mind was, Can we restore areas that have suffered from our abuse? If so, how close to pristine can we get? If finding pristine places was traveling back in time, now we needed to travel to the future. And I found the future in the beloved barren sea of my childhood.

A Pristine Mediterranean?

The Illes Medes (Medes Islands) are a collection of chunks of limestone that glided over slippery layers of rock from the Pyrenees to the Costa Brava off Catalonia more than a hundred million years ago. This geologic anomaly was a gift from the Earth goddess, for eons of wave action have eroded this limestone beautifully, sculpting caves and crevices and creating boulders galore. It's the perfect habitat for fish.

I first saw the Medes Islands as a child, from the beach, one kilometer away. They seemed inaccessible. I thought Poseidon must have thrown those rocks out there, angry over another god's malfeasance. Ten years later I was on a motorboat with my dive gear on below the gray cliffs of those islands, and they looked just as fearsome. I did a backflip and entered the Mediterranean.

A black belt of mussels lined the rock, marking the entrance of the wall into the sea. The wall cut the sea and plunged as deep as I could see. Half-meter-long gilded sea breams (Cousteau's *daurades*) were crunching on the mussels, sending pieces of shell-like black confetti to the bottom. Below the mussels, a skirt of red algae; beneath that, more seaweed of brown and pastel colors, undulating softly with the gentle Mediterranean swell, interspersed with red sea squirts and black sponges. Twenty meters below, blue-and-yellow trees—sea fans—clung onto the rock. I dived down to observe them up close and turned my light on. The light had a magical effect, turning the blues into bright reds. Because warm colors are the first colors lost as light penetrates and is absorbed by seawater, everything below 12 meters looks bluish. A light brings out their true colors.

My old friends Josep Maria Llenas and Miquel Sans—the men who taught me to scuba dive—were diving with me. Josep Maria pointed behind me. I turned around to see a large Mediterranean dusky grouper trying to eat some of my bubbles. It got so close to my face that it covered half my sight. It had a dark brown color with light blotches spread across its entire body, large inexpressive eyes, and thick lips, with a jaw pointing down, creating a facial expression somewhere between sad and indifferent.

Groupers have been annihilated from Mediterranean waters in the last century. Their only sin is that they taste delicious. This grouper, however, was not only unafraid of us but definitely interested in making our acquaintance. Sans was filming us, and as soon as he turned the lights of his underwater camera on, the grouper became more interested in his camera dome than in me. What a narcissistic fish, looking at his own reflection on the glass dome.

Above us appeared a handful of sea basses, the wild type of the farmed *branzino* that is so popular in American seafood restaurants. They were between 60 and 90 centimeters long, shaped like torpedoes, and as they swam, their scales glittered like silver. All of a sudden the torpedoes launched, striking a school of bogues, small sea breams, which dispersed in a second. The sea basses moved so fast that they left a wake behind, making even Usain Bolt look slow. Amid the brown algae an octopus made use of his dynamically colorful cloak, mimicking the background perfectly as he moved across it. Being invisible is essential if you are the favorite snack of groupers.

The Economics of Marine Reserves

Diving at the Medes Islands was like entering one of the films that Jacques Cousteau made in the late 1940s, only this time the film was in color. The abundance I was finding there was not there when I was a child growing up on the coast nearby, because for the last decade the Medes Islands had been declared a marine reserve where fishing was prohibited. Ten years after the reserve was created, fish abundance had increased fivefold. I cannot think of a company share with a 500 percent return in less than ten years that was not a Ponzi scheme. But ocean life, if given a chance, produces extraordinary returns.

**A TRITON SNAIL IN
AN ALGAL FOREST**
Hunted for food and orna-
mental purposes, triton
snail populations have been
decimated throughout the
Mediterranean. In Spain's
Parque Nacional Marítimo
Terrestre del Archipiélago
de Cabrera (Cabrera Archipelago Maritime-Terrestrial
National Park) they are
protected from exploitation,
and the snails roam the reefs
freely once again. (page 222)

A RELIC REEF
Cuba's Jardines de la Reina
National Park contains the
best preserved elkhorn coral
stand in the Caribbean, an
abundant habitat for many
reef fish.

No-take marine reserves are savings accounts we establish in the natural world with a principal set aside that produces interest we can enjoy for a long time to come. When we don't kill fish, they take a longer time to die—science only confirms what's common sense. And as they grow larger, they produce disproportionally more eggs. Some of that production surplus migrates outside of the reserves. Eggs and larvae are dispersed by the ocean currents, and some adult fish move beyond the boundaries of the reserves, too. That spillover helps local fishermen, who tend to be better off fishing around a marine reserve.

My friend Miquel Sacanell, a biologist turned fisherman who fishes on the periphery of the Reserva Marina de les Illes Medes (Medes Islands Marine Reserve), told me that he would not be fishing if the reserve weren't there. Without reserves, the ocean is like a bank account where everybody withdraws but nobody makes a deposit. If we managed fishing smartly, we would not be depleting the ocean's principal. Sadly, most fisheries worldwide are dismally managed.

Reserves Mean Better Diving

As fish come back within marine reserves, tourists come in. The Medes Islands Marine Reserve, less than one square kilometer in size, supports 60,000 sport dives per year. That's the largest concentration of divers anywhere in the Mediterranean, and probably in most of the world. In addition, many thousands more visit the islands with small boats, glass-bottom boats, and kayaks.

Josep Capellà, a local consultant, analyzed the economics of the Medes Islands and found that the reserve has created more than 200 new jobs directly, and that it brings ten million euros per year to the local economy—without accounting for secondary economic benefits. That's 50 times more than the revenue produced by fishing an area 10 times larger than the reserve. It does not take a Nobel Prize in economics to figure out

what model is more profitable, ecologically and economically: reserve versus no protection. Clearly there is a huge appetite for this type of reserve. The fact that 60,000 dives are conducted in this reserve every year does not mean that there are too many divers, but that there are too few reserves.

The Future of Our Ocean

The future of the ocean will be brighter with more marine reserves. They provide returns and insurance. And the science shows that they are more likely to recover after warming or hurricanes or pollution hit them. If we can bring back some of the riches of the Mediterranean—historically, the most overfished sea on the planet—we can certainly restore other places. Reserves are not the only tool to restore the health and productivity of the ocean, but we know that, if well managed, they work extraordinarily well, for marine life and for people.

At the moment of writing this book, only one percent of the ocean was fully protected in no-take marine reserves. Pristine Seas has inspired the protection of 2.2 million square kilometers—about a third of the global no-take area—and more large reserves are coming in the next few years. How much more reserved ocean area do we need? The United Nations' target is to protect 10 percent of the ocean by 2020. Scientific studies suggest that at least 20 percent should be protected. I think it would be fair to protect 50 percent of the ocean—that is, to give back to the ocean half of it, while we manage our activities in the other half with more care.

We cannot destroy in the next few decades what evolution took millions of years to build. We know that marine life does a much better job managing and replenishing itself than we do. It's time to let the ocean restore itself, and it's time for us to become good stewards of the ocean planet we've inherited. ▪

NORTHWESTERN MEDITERRANEAN

Northwestern Mediterranean

Europe
Asia
Africa
ATLANTIC OCEAN

ITALY

FRANCE

LIGURIAN SEA

Genova (Genoa)

Nice
MONACO

Toulon

Mercator Projection
SCALE 1:3,000,000
1 CENTIMETER = 30.0 KILOMETERS; 1 INCH = 47.4 STATUTE MILES

0 25 50 75 100
KILOMETERS

0 25 50 75 100
STATUTE MILES

0 25 50 75 100
NAUTICAL MILES

MONTGRI, MEDES ISLANDS AND BAIX TER NATURAL PARK

FPA   Fully protected area
PPA   Partially protected area

MONTGRI

PPA

L'Estartit

Punta Salinas

Punta Barra   PPA

ILLES MEDES (MEDES ISLANDS)
FPA
Meda Gran

BAIX TER

Meda Petita (Meda Xica)   Carall Bernat

PPA

MEDES ISLANDS MARINE RESERVE

42° 04' N
42° 02' N
3° 12' E
3° 14' E

Ter

Mercator Projection
SCALE 1:150,000
1 CENTIMETER = 1.5 KM; 1 INCH = 2.4 MI

0 1 2 KILOMETERS
0 1 2 STATUTE MILES

PYRENEES

Andorra la Vella
ANDORRA

SPAIN

Cabo Creus

Golf de Roses

L'Escala
Torroella de Montgr
Girona

Costa Brava

Golfe du Lion

42°N

SALINES D'EIVISSA AND FORMENTERA NATURAL PARK

Eivissa (Ibiza)

EIVISSA (IBIZA)

Punta des Jondal

PPA

Punta de la Rama
Isla des Penjats
Punta de sa Torre de ses Portes

S'Espalmador

PPA

FPA
S'Espardell

PPA

S'ESPARDELL SPECIALLY PROTECTED AREA

FPA   Fully protected area
PPA   Partially protected area

PPA   PPA

La Savina

Estany Pudent

FORMENTERA

38° 50' N
38° 40' N
1° 20' E
1° 30' E

Mercator Projection
SCALE 1:600,000
1 CENTIMETER = 6.0 KM; 1 INCH = 9.5 MI

0 5 10 KILOMETERS
0 5 10 STATUTE MILES

SCANDOLA NATURE RESERVE

CMZ   Core Marine Zone
MBZ   Marine Buffer Zone
TZ   Terrestrial Zone

Anse d'Elpa Nera   MBZ

Cala Scandola
Punta Palazzu   Punta Nera
Ilot Palazzu
CMZ
Ile de Gargalu
Baie d'Elbo
Baie de Focolara

MBZ

SCANDOLA NATURE RESERVE

Baie de Solana

TZ

MBZ

Punta Muchillina
Punta Scandola

Golfe de Girolata

Girolata

42° 24' N
42° 20' N
8° 32' E
8° 36' E

Mercator Projection
SCALE 1:250,000
1 CENTIMETER = 2.5 KM; 1 INCH = 4.0 MI

0 2 4 KILOMETERS
0 2 4 STATUTE MILES

Calvi
Galeria
Girolata
Porto

CORSICA
France

Ajaccio

Strait of Bonifacio

Sassari

SARDINIA
Italy

Cagliari

CABRERA ARCHIPELAGO MARITIME-TERRESTRIAL NATIONAL PARK

MU
na Foradada
na Pobra   na Plana

Illa des Conills

MU
RU
na Redona

Cala Santa Maria   RU
MR
Cap de Llebeig   Cap Ventós
SU
MR   MR
l'Olla
RU
MR
CABRERA

RU
MR

Punta de l'Enciola
MR
RA
MR
MU

RA

39° 12' N
39° 08' N
2° 52' E   2° 56' E   3° 00' E

Mercator Projection
SCALE 1:300,000
1 CENTIMETER = 3.0 KM; 1 INCH = 4.7 MI

0 2 4 KILOMETERS
0 2 4 STATUTE MILES

SU   Special use
MU   Moderate use
RU   Restricted use
MR   Marine Reserve
RA   Regeneration area (restricted use)

BALEARIC SEA

Menorca (Minorca)

Palma de Mallorca
Mallorca (Majorca)
Ses Salines

Ibiza (Eivissa, Ivisa)
Eivissa
La Savina
Formentera

BALEARIC ISLANDS

MEDITERRANEAN SEA

MAP KEY

Depth in meters
Sea Level
100
250
500
1000
1500
2000
2500
3000
3500
4000
4500
4600

44°N
42°N
40°N
2°E 4°E 6°E 8°E

THE MEDITERRANEAN SEA has been exploited since antiquity, causing a dramatic depletion of many species, such as the red coral, sharks, and the Mediterranean monk seal. Despite this history of overexploitation, the Mediterranean offers fine examples of marine protection. Modern Mediterranean no-take marine reserves created as early as 1975, although small, have become some of the greatest successes in marine conservation worldwide. These reserves have increased marine life five to ten times relative to unprotected areas, helped replenish nearby stocks of commercial species, and led to the creation of jobs and new income from ecotourism.

WAVING WITH DEEP OCEAN CURRENTS
In an underwater cave in Cabrera Archipelago Maritime-Terrestrial National Park, tube anemones attach to an ocean bottom made of the skeletons of marine animals such as bryozoans and sea urchins.

KALEIDOSCOPIC UNDERWATER COLORS
These small fish, called ornate wrasses, prey on various invertebrates in the miniature algal forests of the Mediterranean. And then they in turn become the prey of larger fish such as groupers—all part of a marine reserve's complex food web with five working levels. *(following pages)*

"In the Medes Islands, I once dived among a multitude of beautiful *daurades;* twenty-five years later I returned to find a nearly empty shooting gallery, still frequented by unrelenting daily squadrons of spearfishermen."

—JACQUES-YVES COUSTEAU

BIG HOPE FOR THE MEDITERRANEAN

Pierre-Yves Cousteau, youngest son of Jacques-Yves Cousteau, observes a Mediterranean dusky grouper at Cabrera Archipelago Maritime-Terrestrial National Park. Protected areas like Cabrera allow large fish to come back and attract divers, who cannot see such large fish anywhere else in the Mediterranean.

LAYING EGGS ON THE DINING ROOM
The nudibranch *Flabellina affinis* is a sea slug that lives in the Mediterranean. Here, one lays thousands of eggs inside a pink ribbon on the species of hydroid that is its main prey.

CRAWLING ON SPICY FOOD
The Mediterranean sea slug *Dondice banyulensis* moves slowly along the branches of a hydroid. This sea slug eats the polyps of hydroids, including the stinging cells, and stores them in its body as protection from predators.

UP CLOSE AND CRABBY
A European spider crab hops over the camera at Corsica's Scandola Nature Reserve Although they are usually inhabitants of deeper waters, spider crabs migrate to shallow algal forests in the spring to reproduce. They have been driven to extinction in many Mediterranean locales because of overfishing.

The Treasure Chamber

In ancient Greek mythology, Medusa was a Gorgon with snakes for hair. She could turn to stone anyone who looked at her. The Greek hero Perseus killed Medusa and put her writhing head in a sack. He intended to give the head to the King of Seriphos as a wedding present. On his way home, Perseus saw the princess Andromeda chained to a rock, threatened by a sea monster. Perseus killed the monster and freed Andromeda. He put Medusa's head down while he washed his hands, and her blood spilled into the sea, turning into red coral.

Red coral is a colonial animal that grows in various depths of water, from the shallows to as deep as 300 meters down. It grows so slowly that a colony takes 300 years to reach the size of a human hand. The Greeks and other Mediterranean peoples have harvested red coral since antiquity, first using a wooden cross with a weight and ropes attached to it. They dragged the cross over underwater reefs, ripping the red coral from the ocean floor, along with everything else growing and attached nearby. Mediterranean cultures valued red coral as if it were a gemstone. Some believed red coral amulets would bring good luck, secure victory in battle, and ward off evil. The less superstitious still regarded red coral jewelry as the height of fashion.

In recent times, scuba divers have harvested the last of the Mediterranean red coral from the shallows, stripping away the few large colonies that remained. Currently, the only old colonies that survive can be found only below 100 meters in depth, and even they are vanishing under the pressure of deep coral divers, thirsty for the last of the Mediterranean's red gold.

That is, all except in one place: the Réserve Naturel de Scandola (Scandola Nature Reserve) in Corsica, a French island in the Mediterranean. Established in 1975 and named a natural World Heritage site by the United Nations in 1983, the reserve harbors in its offshore waters beds of red coral as brilliant and populous as they must have been in the days when people told the myth of Perseus and Andromeda. These rich red coral beds survived because they were inaccessible, not only to the ancients but to modern divers as well. Located inside an underwater cave and at a depth of 20 meters, they provide us the best baseline of the Mediterranean Sea in the days of Odysseus. This brilliant red coral shows us not only how much we have lost, but also what the future of the Mediterranean could be. ■

RED GOLD OF THE DEEP
Mediterranean red coral thrives in an underwater cave in the Scandola Nature Reserve,
one of only a handful of places where healthy red coral stands can be found in shallow
waters. Elsewhere, throughout the Mediterranean, scuba-accessible depths have been
wiped clean of this coral through years of overexploitation.

JARDINES DE LA REINA, CUBA

MAP KEY

Coral reef

Maritime treaty boundary

Depth in meters
Sea Level
100
250
500
1000
1500
2000
2500
3000
3500
4000
4500
5000
5500
6000
6500
7000
7500
8000
8500
8605

Main map labels

UNITED STATES

ATLANTIC OCEAN

NORTH AMERICA

Mississippi River Delta

FLORIDA

U.S. Bahamas

North America
ATLANTIC OCEAN
Jardines de la Reina
PACIFIC OCEAN
South America

Grand Bahama I.

Abaco Islands

Northwest Providence Channel

Bimini Islands

Berry Is.

Eleuthera I.

Nassau

BAHAMA ISLANDS

Cat I.

FLORIDA PLAIN

Miami

GULF OF MEXICO

CAMPECHE ESCARPMENT

U.S. Mex.

Key West

Dry Tortugas

Florida Keys

STRAITS OF FLORIDA

U.S. Cuba

Bahamas Cay Sal

Cay Sal Bank

Santaren Channel

GREAT BAHAMA BANK

Andros Island

Tongue of the Ocean

Exuma Sound

Exuma Cays

San Salvador (Watling)

Rum Cay

Long Island

Great Exuma

Crooked Island Passage

Crooked I.

Acklins Island

Columbus Bank

Mayaguana I.

TROPIC OF CANCER

Arrecife Alacránes (Scorpion Reef)

CAMPECHE BANK

Cabo Catoche

YUCATÁN CHANNEL

Cuba Mex.

Bahía Guadiana

Cabo de San Antonio

Bahía de Corrientes

Arch. de los Colorados

La Habana (Havana)

Golfo de Batabanó

Isla de la Juventud (Isle of Youth)

Arch. de los Canarreos

Arch. de Sabana

Arch. de Camagüey

CUBA

B. de Jigüey

Magallanes Bank

Old Bahama Chan.

Golfo de Ana María

Jardines de la Reina

Santa Cruz del Sur

Golfo de Guacanayabo

Little Inagua I.

Great Inagua Island

Caicos Passage

Grand Caicos

Caicos Islands

TURKS & CAICOS ISLANDS U.K.

Turks Is.

Grand Turk

Mouchoir Bank

Silver Bank

Navidad Bank

HISPANIOLA BASIN

Mérida

Cancún

Isla Cozumel

YUCATÁN PENINSULA

YUCATÁN ESCARPMENT

YUCATÁN PLAIN

YUCATÁN BASIN

GUANCANAYABO TROUGH

Little Cayman

Grand Cayman

George Town

CAYMAN ISLANDS U.K.

Cayman Brac

Santiago de Cuba

Bahía de Guantánamo

U.S. Naval Base Guantanamo Bay U.S.

Golfo de la Gonâve

HAITI

HISPANIOLA

DOMINICAN REPUBLIC

Santo Domingo

Santo Domingo Basin

PUERTO RICO TRENCH

PUERTO RICO U.S.

Misteriosa Bank

CAYMAN TRENCH

CAYMAN TROUGH

Swan Trough

Honduras U.K.

Cuba Jam.

Île de la Gonâve

Pòtoprens (Port-au-Prince)

Windward Passage

Cuba Jam.

Formigas Reef

Navassa I. U.S.

Isla Mona

Mona Passage

MUERTOS TROUGH

BELIZE

Belmopan

GUATEMALA

Turneffe Is.

Northern Caye

Banco Chinchorro

Islas Santanilla (Swan Is.) Honduras

Swan Trough

JAMAICA

Kingston

Pedro Cays

Morant Trough

Albatross Bank

Morant Cays

CARIBBEAN SEA

BEATA RIDGE

HESS ESCARPMENT

Jam. Col.

VENEZUELAN BASIN

COLOMBIAN BASIN

LESSER ANTILLES

Aruba Neth.

Oranjestad

Curaçao Neth.

Bonaire Neth.

Willemstad

Islas Los Roques

RANCHERIA BASIN

MAGDALENA FAN

Pen. de la Guajira

Pen. de Paraguaná

Golfo de Venezuela

Barranquilla

Cartagena

Maracaibo

Lago de Maracaibo

Caracas

COLOMBIA

VENEZUELA

SOUTH AMERICA

Miller Cylindrical Projection

SCALE 1:12,000,000
1 CENTIMETER = 120.0 KILOMETERS; 1 INCH = 189.4 STATUTE MILES

0 100 200 300 400
KILOMETERS

0 100 200 300 400
STATUTE MILES

0 100 200 300 400
NAUTICAL MILES

Inset map

CAYOS ANA MARÍA

Cayo Tío Joaquín

Cayo Arenas

Cayo Burgao

Golfo de Ana María

Arrecife Médanos de la Vela

Punta Blanca

Embarcadero de Santa María

Ensenada de Santa María

Punta Manzanillo

Cayos Paloma

Cayo Santa María

Vertientes

El Congo

Cayo Rabihorcado

Canal de Bretón

Cayo Bretón

CAYOS CINCO BALAS

Laberinto de las Doce Leguas

ARCHIPIÉLAGO DE LOS JARDINES DE LA REINA

Canal Boca Grande

Cayo Grande

CAYOS DE LAS DOCE LEGUAS

Cayo Caballones

Canal de Caballones

CAYOS ANCLITAS

JARDINES DE LA REINA NATIONAL PARK

CARIBBEAN SEA

Cayo Cuervo

Cayos Manuel Gómez

Cayo Algodón Grande

Cayo Algodoncito

Punta Potrerillo

Punta Macurijes

Cayo Cargado

Cayos Bergantines

Cayo Mosquitos

Cayo Bonito

Cayos Gitanos

Cayo Malabrigo

Laguna Lamar

Cayo Hormigas

Cayo Chocolate

CAYOS PINGÜES

Cayos Orihuelo

Cayo Largo

Cayo Cataubo

CAYOS INDIO

Cayo Piedra Grande

Cayo Cachiboca

CAYOS LABERINTO DE LAS DOCE LEGUAS

Cayo Carabinero

Cayo Juan Grin

Cayo Boca Seca

Cayo Campo Santo

Cayo Las Caguamas

Cayos Pilón

Cayo Cabeza del Este

Canal de Cabeza del Este

Mercator Projection

SCALE 1:1,500,000
1 CENTIMETER = 15.0 KILOMETERS; 1 INCH = 23.7 STATUTE MILES

0 5 10 15 20 25 KILOMETERS

0 5 10 15 20 25 STATUTE MILES

CUBA includes not only the main island but also surrounding archipelagoes of islands between the Gulf of Mexico and the Atlantic Ocean. Cuba's waters are the jewel of the Caribbean. Coral reefs, mangroves, and coastal lagoons surround the island, and they support a diverse fauna, including two species of marine crocodiles, sharks, and large groupers. The critically endangered hawksbill turtle nests on the island's remote beaches. Parque Nacional Jardines de la Reina (Jardines de la Reina National Park) is located 100 kilometers south of the island and encompasses more than 600 uninhabited small islands and cays. Some of the last healthy elkhorn coral stands are found here, alongside sea grass beds, mangrove forests, and underwater walls and caves.

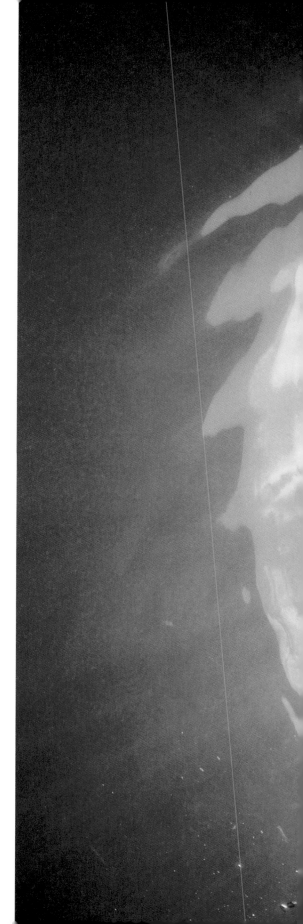

TOP PREDATORS AT WORK
An American crocodile swims in a protected channel between mangroves at Jardines de la Reina National Park, Cuba. With no natural predators, these animals feast on fish, including large tarpon, in the rich waters of the marine reserve.

PROTECTED ABUNDANCE
Jardines de la Reina National Park in Cuba has the largest fish biomass in the Caribbean, including large schools of grunts and snappers. This abundance is rare, however: Less than one percent of Caribbean waters are fully protected from fishing.

WHERE GOBIES GATHER
A golden neon goby stands out against the closed polyps of a great star coral in Cuba. Reef corals in the Caribbean used to cover up to 80 percent of the ocean bottom. Now they have declined to only 5 to 10 percent, a result of human impacts.

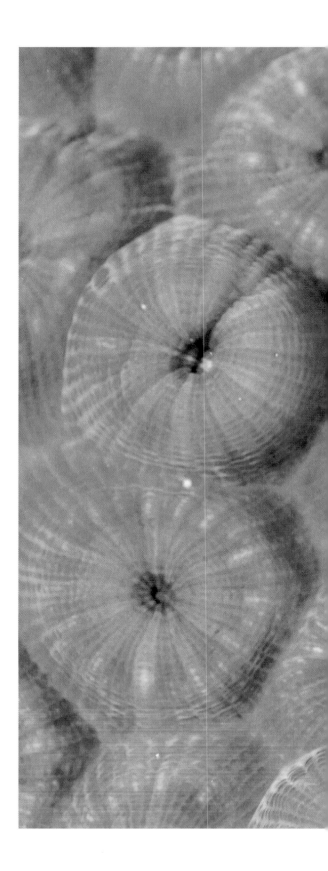

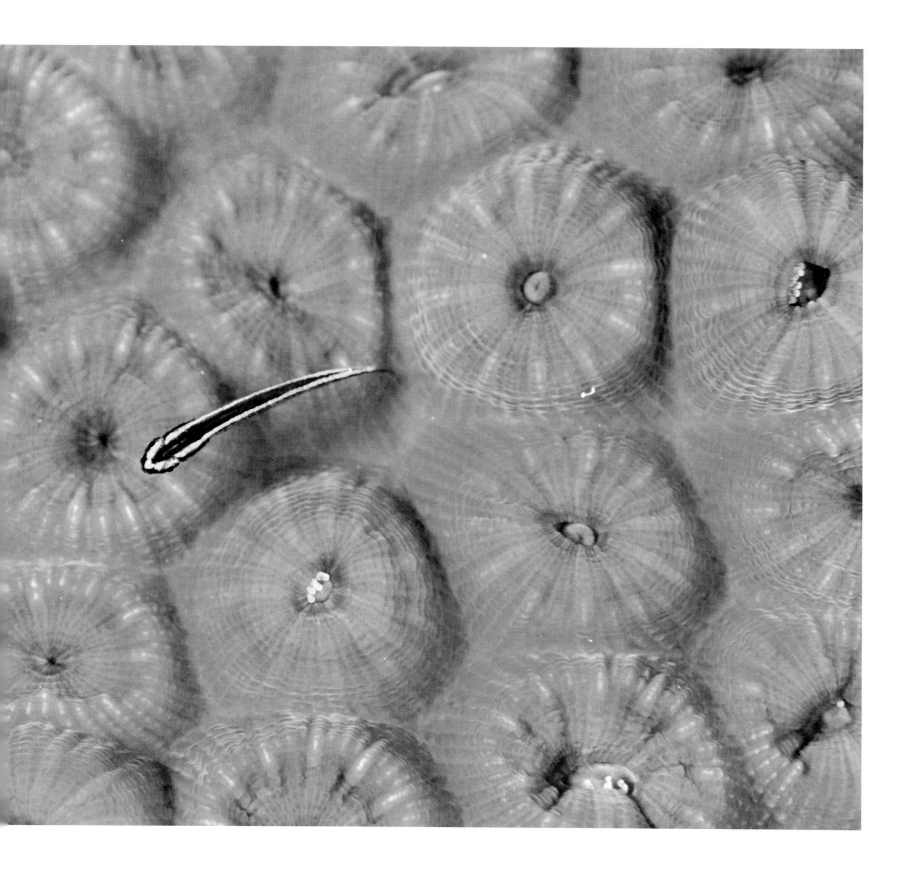

SHARK-FILLED GARDENS
Caribbean reef sharks are abundant in Jardines de la Reina National Park, a no-take area south of Cuba.

UNDERWATER CLEANING ARRANGEMENTS
The Pederson cleaner shrimp serves fish as they pass by, ridding them of parasites as they pause at an anemone cleaning station. Recent research shows that fish visit cleaning stations more often as the size of the host anemone increases. *(following pages)*

"Our task must be to free ourselves . . . by widening our circle of compassion to embrace all living creatures and the whole of nature and its beauty."

—ALBERT EINSTEIN

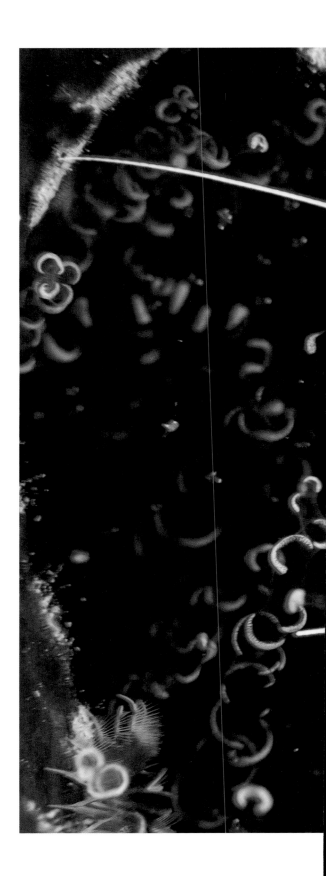

THE RETURN OF THE GROUPERS

The few fully protected areas in the Caribbean have shown how quickly marine life can bounce back. Ninety-kilogram goliath groupers and black groupers (in the background) are a common sight in Cuba's Jardines de la Reina National Park, where fishing is prohibited. Goliath groupers are avid eaters of spiny lobsters and slipper lobsters.

Cuba's Secret Reserve

It was hot, and I was sweating inside my wet suit when we arrived at our destination. We moored our skiff to a small buoy. I stood up and looked around. One minute later, a shadow loomed below the water's surface.

"Here they come!" said Noel López, our diving guide, with a grin. "Six silky sharks! Wait, *eight* silky sharks!" he said.

Today, most people shoot out of the water when anyone claims that sharks are sighted. In contrast, our Pristine Seas team does the opposite. When we hear "Sharks!" we jump *into* the sea as fast as we can. We have learned that an abundance of sharks indicates a healthy ocean area, possibly pristine waters.

And so we jumped.

We were at Jardines de la Reina—the "gardens of the queen," a 240-kilometer-long archipelago located 100 kilometers off the southeastern coast of Cuba. Christopher Columbus first named these islands in honor of Queen Isabella. Uninhabited except for a small tourist operation, the sandy mangrove islands, the sea grass beds, and the reefs around them are off-limits to commercial fishing. This is a 2,170-square-kilometer national park, the largest protected area in Cuba.

The elegant silky sharks circled us as we dove deeper. Once we reached the bottom, our escort shifted to fatter Caribbean reef sharks. Underwater life was amazing and abundant. I made a list of all the fish species larger than 60 centimeters that we saw: In addition to the sharks, we saw goliath grouper, black grouper, Nassau grouper, cubera snapper, and tarpon. This is a Caribbean diver's Santa Claus list—remarkable fish, and species that have been decimated elsewhere, throughout the Caribbean. If we dived randomly across the region, it would take hundreds of dives to see all these species. Here, it took 30 minutes. It was as if we were in a time machine, traveling to Christopher Columbus's Caribbean. We were diving the most pristine oasis left in a sea of overfishing.

Fidel Castro, an avid diver and fisherman, favored the waters at Jardines de la Reina himself. His government realized the potential for high-end tourism and allowed a small diving operation and a strictly regulated catch-and-release fly-fishing operation there. The high tourism revenue helps offset the costs of managing and monitoring the area, making it a prime example of the profitable economic advantage of keeping a natural jewel intact. Jardines de la Reina is a bright spot that gives us hope for the future of the ocean. ∎

A SCHOOL OF SILVER TORPEDOES
Meter-long tarpon swim past the photographer in a flash. Tarpon are a coveted prize for anglers, and at the Jardines de la Reina no-take area they can be commonly seen in large schools.

Epilogue

Not even in my wildest dreams could I have imagined that in my lifetime, I would have the opportunity to explore all these wild places in the ocean, from the Arctic to the tropics. But still harder to believe is that our small, dedicated team and a terrific group of partners have inspired leaders to protect almost as much ocean as Alaska, Texas, and California combined—and in only seven years. I was born too late to fulfill my childhood dream to be a diver on Jacques Cousteau's team, but that motivation made me live the dream in my own way. Our first expedition to a pristine reef was in 2005. In 2008 I proposed the Pristine Seas idea, which the National Geographic Society adopted as a flagship ocean conservation project. In 2009 we helped to inspire the protection of the first pristine site on our list.

When I think I'm dreaming big, someone reminds me that I'm not dreaming big enough. I remember explaining our success stories to a group of friends in 2013, and they started asking very simple questions. How many more pristine places are in the ocean? How much of the ocean is that? Why not try and get *all of them* protected now? What would be the business model for that?

After five years of proving our model for change successful, we decided to scale up. We recruited a top group of partners and sponsors and grew our elite team of explorers, scientists, and communications and policy experts. In 2015 former president Bill Clinton announced our commitment to helping to protect 20 of the wildest places in the ocean by 2019. In other words, we planned to put ourselves out of business in five years. Why? Because it is urgently necessary to save the last pristine places now, before they vanish under human exploitation.

Diving in degraded areas, where fish are small and far apart and the water is filthy with bad microbes, makes me sad and depressed. Pristine places, in contrast, give me hope. They take me back to that childhood state of permanent awe and wonder, to a time when everything looked awesome and full of possibilities. Think of the first time you saw the ocean, the first time you went to the beach with your family, the first time you looked at that blue infinity and dipped your toes in the chilly water. You probably giggled then—and you are probably smiling now.

Pristine ocean ecosystems are the only baselines we have left, the best proxy for what the ocean used to be. They can show us what we've lost through exploitation of the world's natural resources. But most important, they can show us what we can gain—what the future ocean could be. I hope that country leaders will follow the example of the heroes who have already protected some of these pristine places, so that we will live in a world with a healthy ocean that continues to provide, an ocean that is essential to our well-being—and to that of the rest of the inhabitants of our blue planet. ∎

THE MASTER OF THE LAGOON
Juvenile blacktip reef sharks teem in the pristine lagoon of Millennium Atoll,
where they prey on abundant fish safe from larger sharks on the outer reef.

Explorer's Note

We usually start planning expeditions one year in advance. We find and charter a ship; assemble our science team, including local scientists; obtain research and filming permits; gauge the government's interest in a Pristine Seas expedition; and confer with any conservation organizations working in the area. We have to plan far ahead to ship the expedition gear to the closest port from where we will start our expedition.

We spend two to six weeks in each expedition location. In most places we follow up with a trip several months later to show the country's government—and the local population, if there is any—what we found.

The goal of each expedition is to document what is there. That means observing and estimating the abundance and biomass (a measure of the weight of every species per square meter) for as many species of alga, coral, other invertebrates, and fish as possible. That's hundreds of species, and we obtain measures of abundance for each one. When we add microbes, it's tens of thousands of species. And then we film everything that happens, topside and underwater.

This is a normal day on an expedition: We wake up at 7 a.m., have breakfast, and jump into our small boats to go diving around 8:30 a.m. We have a boat for the science team, a boat for the filming team, and another boat for deploying our deep drop-cams and offshore baited cameras.

The scientists conduct two morning scuba dives to count organisms, and the filming team conducts a longer dive with rebreathers. Both teams come back to the mother ship for lunch, when they also refill tanks and recharge batteries. After lunch, both teams dive again. When they come back to the ship, they enter data, download the footage and photos onto hard drives, recharge batteries, fill tanks, and get all the gear ready for the morning after. Meanwhile, the deep-offshore team spends all day out, dropping cameras that film deep-sea habitats—and what's attracted to baited containers near the surface—and returns shortly before dinner.

At dinner there is always a lively discussion: What we accomplished that day, any surprises we encountered, and our work plan for the day after, which shapes the route planned by the expedition leader and ship's captain after dinner. After dinner we enter the day's data into our computers, back up all images, and write our expedition journals. The expedition leader writes the day's blog for the National Geographic website and sends it via satellite.

Having done all that, we retire to our cabins and collapse, with the satisfaction of good work done. But at least once during an expedition we watch a film after dinner. Our favorite, watched over and over, is Wes Anderson's *Life Aquatic With Steve Zissou*.

We repeat this routine daily, as long as the expedition lasts. It is exhausting work, but we could not be happier to have the sea as our office. ■

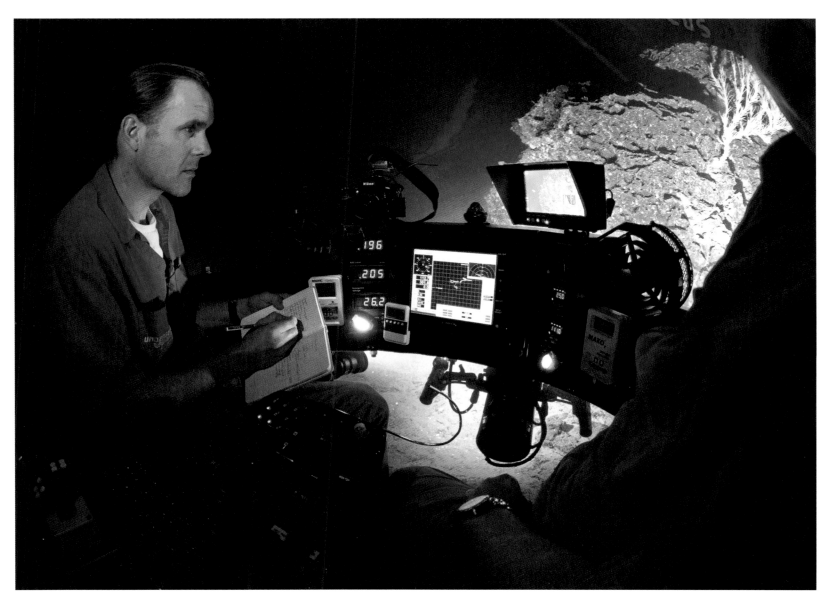

SHALLOW TO DEEP EXPLORATION
Enric Sala documents the marine life found during a trip to the deep waters
off the Desventuradas, Chile, in February 2013.

Finding Pristine Seas

hat are the most pristine places in the ocean? To answer that question, in 2008 I sat over a National Geographic map of the world and circled the most remote places in the ocean: high in the Arctic, around Antarctica, islands in the middle of the ocean. An online search showed there was little scientific information about most of these places. A few had not even been surveyed by marine biologists before.

I recruited my colleagues William Cheung of the University of British Columbia and Ben Halpern of the University of California, Santa Barbara, to conduct a rigorous analysis of the best preserved places in the ocean. We compiled global maps of human threats to ocean life, including the impact of fishing, land-based sources of pollution, and shipping lanes, among others (shown here). From those we obtained a composite showing the gradient from the highest human impact (red) to lower impact (yellow) (top map, page 264).

From there, we decided to focus our efforts primarily on areas within the 200-mile exclusive economic zone of countries—that is, portions of ocean under the jurisdiction of specific countries, thanks to a 1982 United Nations agreement—because in our experience, working with individual countries

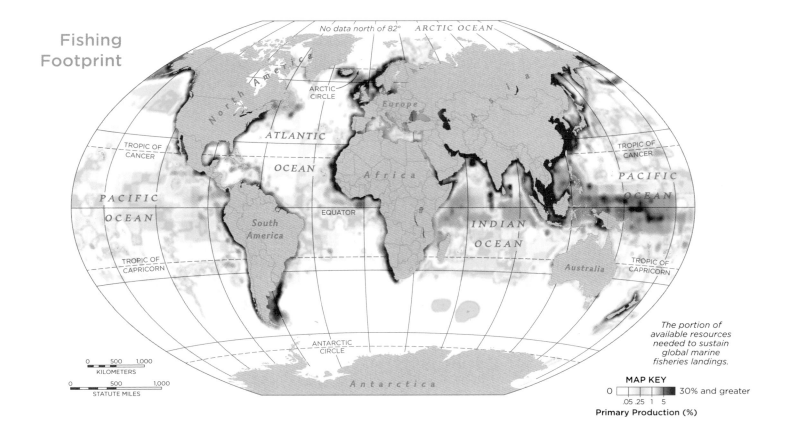

Fishing Footprint

The portion of available resources needed to sustain global marine fisheries landings.

MAP KEY

0 — 30% and greater
.05 .25 1 5
Primary Production (%)

Ocean-Based Pollution

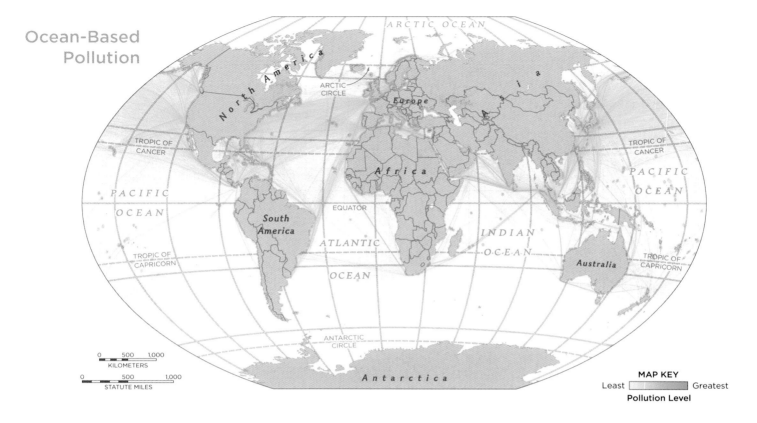

ARCTIC OCEAN

North America

ARCTIC CIRCLE

Europe

Asia

TROPIC OF CANCER

TROPIC OF CANCER

PACIFIC OCEAN

PACIFIC OCEAN

Africa

South America

ATLANTIC OCEAN

EQUATOR

INDIAN OCEAN

Australia

TROPIC OF CAPRICORN

TROPIC OF CAPRICORN

ANTARCTIC CIRCLE

Antarctica

0 500 1,000
KILOMETERS

0 500 1,000
STATUTE MILES

MAP KEY

Least ▨▨▨▨ Greatest

Pollution Level

Coastal Pollution

ARCTIC OCEAN

North America

ARCTIC CIRCLE

Europe

Asia

ATLANTIC OCEAN

TROPIC OF CANCER

TROPIC OF CANCER

PACIFIC OCEAN

PACIFIC OCEAN

Africa

OCEAN

South America

EQUATOR

INDIAN OCEAN

Australia

TROPIC OF CAPRICORN

TROPIC OF CAPRICORN

ANTARCTIC CIRCLE

Antarctica

0 500 1,000
KILOMETERS

0 500 1,000
STATUTE MILES

MAP KEY

Less polluted ▨▨▨▨ More polluted

Human Impact

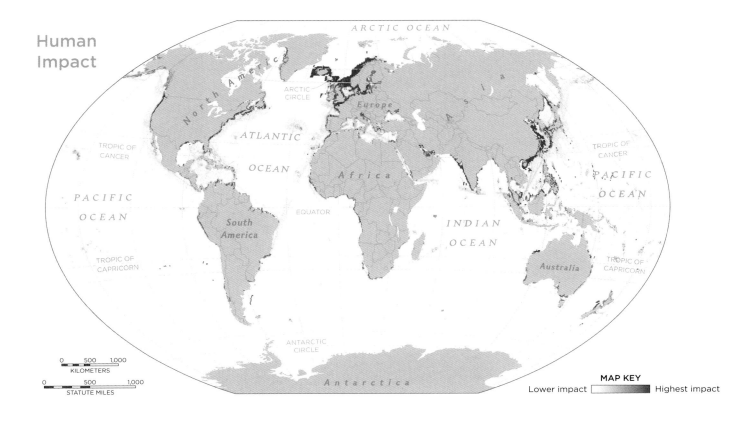

ARCTIC OCEAN

North America

ATLANTIC OCEAN

ARCTIC CIRCLE

Europe

Asia

TROPIC OF CANCER

Africa

PACIFIC OCEAN

PACIFIC OCEAN

EQUATOR

South America

INDIAN OCEAN

TROPIC OF CANCER

TROPIC OF CAPRICORN

Australia

TROPIC OF CAPRICORN

ANTARCTIC CIRCLE

Antarctica

0   500   1,000
KILOMETERS

0   500   1,000
STATUTE MILES

MAP KEY

Lower impact ▢ Highest impact

Extent of Sea Ice

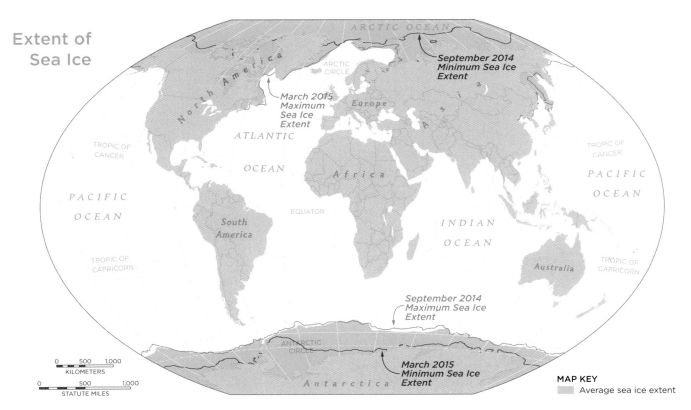

ARCTIC OCEAN

North America

September 2014 Minimum Sea Ice Extent

ARCTIC CIRCLE

March 2015 Maximum Sea Ice Extent

Europe

Asia

ATLANTIC OCEAN

TROPIC OF CANCER

Africa

PACIFIC OCEAN

PACIFIC OCEAN

EQUATOR

South America

INDIAN OCEAN

TROPIC OF CANCER

TROPIC OF CAPRICORN

Australia

TROPIC OF CAPRICORN

September 2014 Maximum Sea Ice Extent

ANTARCTIC CIRCLE

March 2015 Minimum Sea Ice Extent

Antarctica

0   500   1,000
KILOMETERS

0   500   1,000
STATUTE MILES

MAP KEY

▨ Average sea ice extent

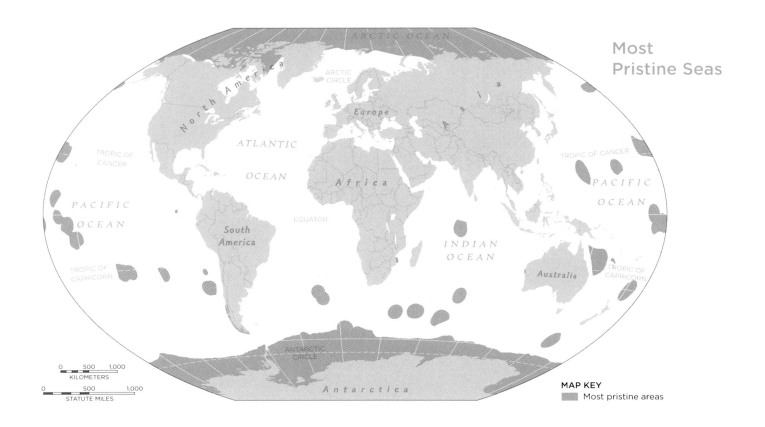

MAP KEY
Most pristine areas

0 500 1,000
KILOMETERS

0 500 1,000
STATUTE MILES

meant that new marine reserves could be created in the short term. The high seas—the waters beyond national jurisdiction—are much more difficult to protect because they require international agreements; to date, only a handful of reserves exist in the high seas. We needed quick wins to show the way for other leaders to follow.

Large areas in the Arctic Ocean and the Antarctic Circle have been covered permanently by sea ice (map opposite), which means that these areas have not been affected by fishing to date. We can safely assume that the waters below the ice have been kept pristine. However, the permanent sea-ice coverage has been decreasing since the 1970s because of global warming, and fishing fleets are starting to move to higher latitudes in search of the last unexploited

fishing grounds. It is therefore urgent to include these areas in any global ocean conservation analysis. We overlapped a map of these areas onto our map of human impacts, and thus identified the areas of the ocean that had experienced the lowest human impact around the world: the most pristine seas on the planet (map above).

These areas account for a little less than 10 percent of the ocean's surface. Most of them are unprotected. And so they represent our targets for the next few years. Together with partners and governments, we will work to help protect as much pristine ocean as possible, so that we can witness a system of marine reserves protecting a fair representation of the ocean's ecosystems during our lifetimes. ▪

Acknowledgments

Our explorations are a team effort, and the most successful I have experienced in my career. What people have done to help Pristine Seas, and the stories we have compiled at every location we have visited, deserve another book, which would likely be very entertaining. Here I want to acknowledge those who contributed the most to the success of Pristine Seas to date. I am deeply grateful to the National Geographic Society, especially Terry Garcia, who believed in the Pristine Seas idea in 2008 and brought me in; and to John Fahey, Gary Knell, and other executives who have since shown continuous support. Thanks to the Science and Exploration team, especially Mark Bauman, Glynnis Breen, Sarah Laskin, Lucie McNeil, and Alex Moen; the Development Office, David Bennett, Jacquie Hollister, Kristin Rechberger, and Bill Warren; National Geographic Studios, National Geographic Channel, NatGeo Wild, Brooke Runnette, Geoff Daniels, and their teams; and *National Geographic* magazine, Chris Johns, Kathy Moran, Oliver Payne, and Sadie Quarrier. A million thanks to the wonderful *National Geographic* Books team, in particular Sanaa Akkach, who created the beautiful design; Susan Hitchcock, who was a terrific editor; Lisa Thomas, who had the vision to make this book possible; and all the other folks listed in the masthead. And thanks to everyone at National Geographic who has helped in any way to make Pristine Seas a success.

Pristine Seas could not have happened without the individuals, foundations, and corporations that provided the needed financial support. I'm deeply indebted to my friend Ted Waitt, who was our first donor and has been a close adviser and wonderful expedition partner. Enormous thanks go to Blancpain (our first and loyal major corporate partner), the Campbell Foundation, the Case Foundation, Davidoff Cool Water, Google, Leonardo DiCaprio Foundation, Roger and Rosemary Enrico Foundation, the Leona M. and Harry B. Helmsley Charitable Trust, Jynwel Foundation, Lindblad Expeditions, Moore Family Foundation, Oracle Education Foundation, Patagonia, Prince Albert II of Monaco Foundation, Philip Stephenson Foundation, Waitt Foundation, and the Walton Family Foundation; and especially to Björk, Scott Burns, Samantha and Keith Campbell, Jean and Steve Case, Yvon Chouinard, John Codey, Bob Cook, Alain Delamuraz, Leonardo DiCaprio, Roger Enrico, Bernard Fautrier, Loic and Philippe Gouzer, Marc Hayek, Nayla Hayek, Nick Hayek, Alex Heath, Jacob James, Alexis Kitsopoulos, Sven Lindblad, Jho Low, Gordon and Steven Moore, Malinda Pennoyer, Rick Ridgeway, Vicki and Roger Sant, Phil Stephenson, and Justin Winters. Thanks to many individual donors who have contributed to Pristine Seas, including Sarah Argyropoulos, Mike Bonsignore, Alice and David Court, George Gund, Iara Lee, Ann Luskey, Mark Moore, and Donna and Garry Weber.

Our extraordinary Board of Advisers made the expansion of Pristine Seas possible. I am indebted to my friends who help us with strategic advice and implementation: H.S.H. Prince Albert II of Monaco, who graciously chairs the board; Samantha Campbell, Steve Case, John Codey, Leonardo DiCaprio, José María Figueres, Terry Garcia, Marc Hayek, Sven Lindblad, Jho Low, Kristin Rechberger, and Ted Waitt.

I am honored and proud to work—and to have worked—with an extraordinary team of passionate people who are obsessed about saving the last wild places in the ocean, the Pristine Seas team: Jose A. Arribas, Kike Ballesteros, Mark Bauman, Maureen Dolan-Galaviz, Kaya Ensor, Tatiana Faramarzi, Mike Fay, Alan Friedlander, Neil Gelinas, Maggie Hines, Kalee Kreider, Nathan Lefevre, Curtis Malarkey, Dave McAloney, Monica Medina, Brian Newell, Scott Ressler, Paul Rose, and Manu San Félix. Thanks to the brilliant engineers at our Remote Imaging Lab: Eric Berkenpas, Brad Henning, Mike Shepard, and Alan Turchik; and to the digital gurus Amy Bucci, Kate Coughlin, Andrew Howley, and Susan Poulton. Thank you to "Nitrox" Bob Olson and Nitrox Solutions for all of your support throughout the years.

I learned a lot from my fellow explorers and photographers, and I thank them for their availability and advice: Bob Ballard, James Cameron, Reza Deghati, David Doubilet, Sylvia Earle, Mike Fay, Beverly and Dereck Joubert, Zafer Kizilkaya, Paul Nicklen, Cory Richards, David de Rothschild, Manu San Félix, Brian Skerry.

Thanks to my colleagues in academia who influenced me the most on my search for pristine places: Kike Ballesteros, Paul Dayton, Alan Friedlander, Jeremy Jackson, Nancy Knowlton, Enrique Macpherson, and Mikel Zabala. I'm grateful to my friends at the forum of the Young Global Leaders at the World Economic Forum for all they taught me about economics, politics, and leadership, and for access to leaders, especially David Aikman, Ida Auken, Cristina Bitar, Jaime de Bourbon Parme, Borge Brende, Katherine Brown, Beth and Angel Cabrera, François P. Champagne, John Dutton, Marco Fiorese, Katherine and Jeremy Garret-Cox, Valerie Keller, Nancy Lublin, John McArthur, Rosy Mondardini, John Nevado, Constanze and Oliver Niedermaier, H.R.H. Haakon and Mette-Marit of Norway, Mina al-Oraibi, Coen van Oostrom, Mabel van Oranje, Yana Peel, Luis G. Plata, Eric Roland, Daniel and Carolina Sachs, Klaus Schwab, Lara Setrakian, Dan Shine, Melanie Walker, and Yan Yanovski. Thanks to the people and organizations that have recognized the work of Pristine Seas and helped us promote the importance of saving the last wild places in the ocean: President Bill Clinton, Gil Grosvenor, Caroline Hermans, Debbie Levin; Clinton Global Initiative, Environmental Media Association, Fundación Amigos de la Isla del Coco, Spanish Geographical Society, the Explorers Club, and the World Economic Forum. Thanks to Wes Anderson and Bill Murray for *The Life Aquatic With Steve Zissou,* which we watch on board during every expedition, to get in the right mood and not be too serious.

Thanks to the following individuals and organizations that made our work possible in countries that harbor pristine places:

Chile

Thanks to the Chilean Navy and its Hydrographic and Oceanographic Service (SHOA), Commission of Maritime Interests, Fisheries and Aquaculture of the Chilean Senate, Defense Ministry of Chile, Fundación Pumalín, the Department of Fisheries National Forest Service of Chile (CONAF), OCEANA, ORCA, the crews of the *Argo* and the Waitt Foundation's research vessel, and the Rapa Nui community at Easter Island. Thank you to President Sebastián Piñera for creating Motu Motiro Hiva Marine Park and to Senator Antonio Horvat for his leadership in ocean conservation in Chile. Thanks to Foreign Minister Heraldo Muñoz for committing to create the Desventuradas Islands marine park by October 2015. We are grateful to MP Enrique Acorssi, Jim Beets, Cristina Bitar, Shmulik Blum, Easter Island governor Carmen Cardinali, Edmundo Edwards, Paula Escobar, Michel and Henry García, Carlos Gaymer, Admiral Edmundo González, Ignacio González, Matthias Gorny, Avi Klapfer, Ximena Muñoz, Yosy Naaman, Uri Pate, Andrés Rodrigo and the crew of the OPV-82 *Comandante Toro,* Eduardo Sorensen, Leonor Varela, Ted Waitt, Lucas Zañartu, Rapa Nui mayor Luz Zasso Paoa, and, last but not least, my dear friend and partner in conservation, Alex Muñoz.

Costa Rica

Thanks to Kyler Abernathy, Octavio Aburto-Oropeza, Randall Arauz, Jenny Ash, Maureen Ballestero, Leo Blanco, Allan Bolaños, Keith Campbell, Colleen Cassity, René Castro, President Laura Chinchilla, Ford Cochran, Jorge Cortés, Sylvia Earle, Muni Figueres, Bill Ginn, Geiner Golfín, Kristen Green, Peter Harren, Nicolás Ibarguen, Jorge Jiménez, Avi Klapfer, Zdenka Piskulich, Marco Quesada, Fernando Quirós, Manuel Ramírez, Sandra Ramírez, Bruce Robison, Michael Rothschild, Fabián Sánchez, Carla Sarquis, Rick Starr, Carlos Uribe, Edith Widder, Brian J. Zgliczynski, and our friends and colleagues at Área de Conservación Marina Isla del Coco (ACMIC), Centro de Investigación en Ciencias del Mar y Limnología (CIMAR) of the Universidad de Costa Rica,

Conservation International, Fundación Amigos de la Isla del Coco, Marviva, Preservación de Tortugas Marinas, Sistema Nacional de Áreas Protegidas (SINAC), the Nature Conservancy, the Undersea Hunter Group, and the crews of the M.V. *Hanse Explorer* and the M.V. *Argo*.

Cuba

Thanks to Archie Carr III, Jean and Steve Case, Mike Bonsignore, Claire Boobyer, Tony Cárdenas, Cole Charlton, Roger Enrico, Chris and Sean Guinness, Noel López, Ann Luskey and the crew of the *Sirenuse,* Pepe Omegna, Fabian Pina, Roger Sant, Mark Spalding, the staff at Avalon Diving Center, National Geographic Expeditions, Ocean Foundation, and Wildlife Conservation Society.

Franz Josef Land

Thanks to H.S.H. Prince Albert II of Monaco, Ann Kristin Balto, David Balton, Susan Barr, Victor Boyarsky, Pete Capelotti, Anastasia Chernobrovina, Alexander Chichaev, Roman Ershov, Bernard Fautrier, Michael Jourdan, Elena Kalinina, Andrey Kameniev, Ambassador Sergey Kislyak, Jens Petter Kollhøj, Irina Kovalevich, Ragnar Kvam, Maria Gavrilo, Yuri Gavrilov, Julia Gourley, Andy Mann, Darya Martynova, Lucie McNeil, Anne Melgård, Charlotte Moore, H.R.H. Haakon and Mette-Marit of Norway, Gennady Oleynik, Børge Ousland, Oliver Payne, Bruno Philipponnat, David Quammen, Sadie Quarrier, Steve Quistad, Cory Richards, Rick Ridgeway, Forest Rohwer, Paul Rose, Defense Minister Sergey Shoigu, Andrey Shorshin, Jan-Gunnar Winther, Guro Tangvald, and all the participants and crew of the 2013 expedition to Franz Josef Land, except those who smoked near the oxygen and fuel tanks. Thanks to the Murmansk Shipping Company, National Geographic Library and Archives, *National Geographic* magazine, National Library of Norway, Norwegian Polar Institute, Patagonia, Royal Geographical Society, Russian Arctic National Park, Russian Geographical Society, Russian Museum for the Arctic and Antarctic, Scott Polar Research Institute, and Smithsonian Air and Space Museum Archives.

Gabon

Thanks to President Ali Bongo Ondimba, Mike Fay, Koumba Koumbila, Ted Waitt, Lee White, Parcs Gabon, Total Gabon, Perenco, Vaalco, and Wildlife Conservation Society for making it possible for us to survey Gabon's waters. Thanks to Richard Grissell, Joe Lepore, and the crew of the Waitt Foundation's research vessel for making our work safe and easy.

Line Islands (United States and Kiribati)

We are grateful to Kiribati president Anote Tong for his continuous support and for protecting the territorial waters of the southern Line Islands, and to Tebao Awerika, Lawrance Bailey, Ivan Gayler, Emilio Gerov, Arjun Gupta, Peter Harren, Jeremy Jackson, Les Kaufman, Zafer Kizilkaya, Michael Lang, Sven Lindblad, Jonathan Littenberg, Ann Luskey, Mark Moore, Forest Rohwer and his microbiology team, John and Tyler Rowe, Shari Sant, Ed Scripps, Jennifer Smith, John Steinitz, Nancy Knowlton, Brian Skerry, Tim Soper, Greg Stone, Tukabu Teroroko, Teuea Toatu, Lisa Trotter, Mike Velings, Ted Waitt, and the scientists participating on our three expeditions to the Line Islands between 2005 and 2009. Thanks to the Fairweather Foundation, Kiribati Ministry of Environment, Lands and Agriculture Development, Medical Foundation for the Study of the Environment, National Geographic Society Committee for Research and Exploration, *National Geographic* magazine, National Geographic Television, Palmyra Atoll Research Station and staff, San Diego State University, Scripps Institution of Oceanography, U.S. Fish and Wildlife Service, the Nature Conservancy, and the crews of the *Hanse Explorer, Searcher,* and *White Holly.* Thanks to Presidents George W. Bush and Barack Obama for creating and expanding, respectively, the Pacific Remote Islands Marine National Monument, and to Mike Boots, Jim Connaughton, Secretary of State John Kerry, Jane Lubchenco, Undersecretary of State Cathy Novelli, John Podesta, Undersecretary of Commerce Kathryn Sullivan, and Sally Yozell, for their work to make it happen.

Mediterranean

Thanks to the people and institutions that made our 2010 expedition to the Mediterranean possible and who joined our adventure: H.S.H. Prince Albert II of Monaco, José Amengual, Gemma Aymerich, Leo Blanco, Josep Clotas, Pierre-Yves and Francine Cousteau, Jean-Marie Dominici, Bernard Fautrier, Joaquim Garrabou, Toni Grau, Jean-Georges Harmelin, Georges Kern, Josep M. and Marta Llenas, Alex Lorente, Enrique Macpherson, Catalina Massuti, Jorge Moreno, Cristina Ozores, Patrice Quesnel and the crew of the *Alcyone,* Miquel Sans, Ted Waitt and the crew of the Waitt Foundation research vessel, Mikel Zabala, Cabrera Archipelago Maritime-Terrestrial National Park, Freus Marine Reserve, IWC Schaffhausen, Marine Resources Service and Biodiversity Department of the Balearic Islands government, Medes Islands Marine Reserve, Corsica Natural Regional Park, Scandola Nature Reserve, the Cousteau Society, and Vellmarí.

New Caledonia

Thank to Eric Brown, Mike Dessner, Mark Downes, Roger Enrico, Hugues Gossuin, Richard Grissell, Pierre Labrosse, Joe Lepore, Laurent Vigliola, Ted Waitt, Laurent Wantiez, Commune d'Ouvéa, Department of Marine Affairs and Fisheries of New Caledonia, government of New Caledonia, Institut de Recherche pour le Développement (IRD), Université de Montpellier 2, Université de Nouvelle Calédonie, and the crew of the Waitt Foundation's research vessel for providing the best conditions for our work.

Palau

Thanks to Tova and Navot Bornovski, Jennifer Caselle, Pat Colin, Yim Golbuu, Marine Gouezo, Seth Horstmeyer, Noah Idechong, Rebluud Kesolei, Tom Letessier, Jessica Meewig, Dawnette Olsudong, Matt Rand, Keobel Sakuma, Nanae Singeo, Steven Victor, Fish 'n Fins, Koror Stare, Ngemelis State, government of Palau, Ollei fishing community, Palau Conservation Society, Palau International Coral Reef Center, Pew Environment Group, the Nature Conservancy, and the crew of the *Ocean Hunter III.* Thanks to President Tommy Remengesau for his support and his inspiring vision of a Palau national marine sanctuary.

Pitcairn Islands

First of all, thanks to the Pitcairn Island community, who hosted us in their homes and showed us the secrets of their island and ocean; their commitment and that of the Pitcairn Council were key to the creation of Pitcairn Islands Marine Reserve in March 2015. Thanks to H.S.H. Prince Albert II of Monaco, Renee Braden, Heather Bradner, Sir Richard Branson, Eric Brown, Jennifer Caselle, Charles Clover, Chuck Cook, Alistair Gammell, Cathy Hunter, Mark Jenkins, Nigel Jolly, Michael Jourdan, Justin Kenney, Sir David King, Justin Mundy, Jay Nelson, Matt Rand, Jo Royle, Elisabeth Whitebread, and Brian J. Zgliczynski. We are indebted to Maitland Political, Mares, National Geographic Archives, Pew Environment Group, Poseidon Diving Systems, Prince Charles's International Sustainability Unit, and the crew of the *Claymore II.* Thanks to British prime minister David Cameron for protecting the waters around the Pitcairn Islands.

And finally, my deepest gratitude goes to my life co-pilot, Kristin Rechberger, my closest adviser and partner, who makes my life extra special. Her support and love are what keep me hoping for a better world.

Credits

Artwork by Mesa Schumacher: pages 21 and 22-23
All photographic images by Enric Sala except:
page 17, Manu San Félix
page 91, Luis Marden/National Geographic Creative

page 209 (UP), Anthony Fiala/National Geographic Creative;
   (LO) Kristin Rechberger
page 261 Avi Klapfer

Map Sources

General Sources
Shaded Relief and Bathymetry
ETOPO1/Amante and Eakins, 2009; GTOPO30, USGS EROS Data Center, 2000; TCarta Marine, LLC.
NASA Jet Propulsion Laboratory, Shuttle Radar Topography Mission (SRTM). Available online at www2.jpl.nasa.gov/srtm.
The International Bathymetric Chart of the Arctic Ocean (IBCAO), Version 3.0.

Protected Areas and Boundaries
IUCN and UNEP-WCMC (2013) The World Database on Protected Areas (WDPA) Cambridge, UK: UNEP-WCMC. Available online at www.protectedplanet.net and mapatlas.org.
VLIZ (2014). Maritime Boundaries Geodatabase, Version 8. Available online at www.marineregions.org.
U.S. Department of State, Humanitarian Information Unit, Large Scale International Boundaries (LSIB). Available online at https://hiu.state.gov/data/data.aspx.

Coral Reefs
Global Distribution of Coral Reefs (2010) Dataset
UNEP-WCMC, WorldFish Centre, WRI, TNC (2010). Global distribution of warm-water coral reefs, compiled from multiple sources including the Millennium Coral Reef Mapping Project. Includes contributions from IMaRS-USF and IRD (2005), IMaRS-USF (2005), and Spalding et al. (2001). Cambridge, UK: UNEP World Conservation Monitoring Centre. Available online at http://data.unep-wcmc.org/datasets/1.

Other
Global Self-consistent Hierarchical High-resolution Geography (GSHHG): NOAA Earth Reference Seamount Catalog. Available online at earthref.org/SBN.
U.S. Board on Geographic Names (BGN). Toponymic information is based on the Geographic Names Data Base, containing official standard names approved by the United States Board on Geographic Names and maintained by the National Geospatial-Intelligence Agency.

Finding Pristine Seas maps (pages 262–265)
Fishing Footprint map
Sea Around Us project (www.seaaroundus.org, updated from Swartz et al. 2010).

Swartz, Wilf, Enric Sala, Sean Tracey, Reg Watson, and Daniel Pauly. 2010. "The Spatial Expansion and Ecological Footprint of Fisheries (1950 to Present)." *PloS one* 5(12): e15143.

Ocean-Based Pollution and Coastal Pollution Sources maps
Halpern, Benjamin S., Shaun Walbridge, Kimberly A. Selkoe, Carrie V. Kappel, Fiorenza Micheli, Caterina D'Agrosa, John F. Bruno, Kenneth S. Casey, Colin Ebert, Helen E. Fox, Rod Fujita, Dennis Heinemann, Hunter S. Lenihan, Elizabeth M. P. Madin, Matthew T. Perry, Elizabeth R. Selig, Mark Spalding, Robert Steneck, and Reg Watson. "A Global Map of Human Impact on Marine Ecosystems." *Science* (2008), 948-952.

Human Impact map
Sala, Enric, William Cheung, and and others. In prep.

Extent of Sea Ice map
Fetterer, F., K. Knowles, W. Meier, and M. Savoie. 2002, updated daily. Sea Ice Index. Ice Extent: September 2014; March 2015. Boulder, Colorado: National Snow and Ice Data Center. Available online at http://dx.doi.org/10.7265/N5QJ7F7W.

Most Pristine Seas map
Sala, Enric, William Cheung, and and others. In prep.

Finding Pristine Seas maps (pages 262–265) data consulting
William Cheung

Join Us in Protecting the Ocean

National Geographic's Pristine Seas is dedicated to the protection of the last wild places in the ocean. There are few places that have not been impacted by humans. These are remote places far from human habitation, without fishing and pollution, places that look like the ocean eons ago. They are key for preserving unique biodiversity and global ocean health, and also serve as a baseline for our conservation efforts.

Initiated and led by Dr. Enric Sala, National Geographic Explorer-in-Residence, the Pristine Seas project is an exploration, research, and media effort to help protect these ocean wildernesses. These pristine places are unknown by all but long-distance fishing fleets, which have started to encroach on them. By combining exploration, scientific and economic research, communications, and governmental negotiations, Pristine Seas helps to inspire country leaders to save these places before they vanish.

As of May 2015, only one percent of the ocean is fully protected. National Geographic's Pristine Seas has contributed to the protection of six key areas: Pacific Remote Islands Marine National Monument, United States (1,270,000 sq km in 2014); Southern Line Islands (9,000 sq km in 2014); Seamounts Marine Managed Area, Costa Rica (10,000 sq km in 2011); Motu Motiro Hiva Marine Park, Chile (150,000 sq km in 2010); a system of marine parks covering 20 percent of Gabon's waters (47,000 sq km in 2014); and Pitcairn Islands Marine Reserve (834,000 sq km in 2015). Marine reserves in progress thanks to Pristine Seas and partners amount to a potential additional area of 2 million square km in tropical and temperate seas as well as the Arctic.

Over the next few years, Pristine Seas is targeting 16 more pristine places that are still unprotected. These areas could add more than 5 million square km, more than doubling the current area that is fully protected globally.

You can join Dr. Sala and his team and become an ocean hero, helping to protect these last wild places in the ocean, supporting the global goal of fully protecting 10 percent of the world's oceans by the year 2020.

For more information or to join the cause, visit *pristineseas.org*.

About the Author

Enric Sala is a marine ecologist and National Geographic Explorer-in-Residence dedicated to restoring the health and productivity of the ocean. He obtained his Ph.D. in ecology from the University of Aix-Marseille, France, in 1996 and worked as professor at the Scripps Institution of Oceanography and at Spain's National Research Council (CSIC) before joining National Geographic. His more than 120 scientific publications are widely recognized and used for real-world conservation efforts such as the creation of marine reserves.

He is a Fellow of the Royal Geographical Society and was recognized as a 2008 Young Global Leader at the World Economic Forum in Davos. In 2013 he received both the Research Award of the Spanish Geographical Society and the Lowell Thomas Award of the Explorers Club. His 2010 TED Talk, "Glimpses of a Pristine Ocean," has been viewed more than a quarter million times online. He serves on a number of advisory boards of international organizations and governments.

Pristine Seas
Enric Sala

Published by the National Geographic Society

Gary E. Knell, *President and Chief Executive Officer*

John M. Fahey, *Chairman of the Board*

Declan Moore, *Chief Media Officer*

Chris Johns, *Chief Content Officer*

Prepared by the Book Division

Hector Sierra, *Senior Vice President and General Manager*

Lisa Thomas, *Senior Vice President and Editorial Director*

Jonathan Halling, *Creative Director*

Marianne R. Koszorus, *Design Director*

Susan Tyler Hitchcock, *Senior Editor*

R. Gary Colbert, *Production Director*

Jennifer A. Thornton, *Director of Managing Editorial*

Susan S. Blair, *Director of Photography*

Meredith C. Wilcox, *Director, Administration and Rights Clearance*

Staff for This Book

Sanaa Akkach, *Art Director*

Matt Propert, *Photo Editor*

Carl Mehler, *Director of Maps*

Gregory Ugiansky, *Map Research and Production Manager*

Matthew W. Chwastyk and Michael McNey, *Contributing Cartographers*

Juan José Valdés, *The Geographer*

Maureen J. Flynn, Michael Fry, Julie A. Ibinson, and Gus Platis, *Map Editors*

Michelle Cassidy, *Editorial Assistant*

Marshall Kiker, *Associate Managing Editor*

Judith Klein, *Senior Production Editor*

Mike Horenstein, *Production Manager*

Katie Olsen, *Design Production Specialist*

Nicole Miller, *Design Production Assistant*

George Bounelis, Manager, *Production Services*

Rahsaan Jackson, *Imaging*

The National Geographic Society is one of the world's largest nonprofit scientific and educational organizations. Founded in 1888 to "increase and diffuse geographic knowledge," the member-supported Society works to inspire people to care about the planet. Through its online community, members can get closer to explorers and photographers, connect with other members around the world, and help make a difference. National Geographic reflects the world through its magazines, television programs, films, music and radio, books, DVDs, maps, exhibitions, live events, school publishing programs, interactive media, and merchandise. *National Geographic* magazine, the Society's official journal, published in English and 38 local-language editions, is read by more than 60 million people each month. The National Geographic Channel reaches 440 million households in 171 countries in 38 languages. National Geographic Digital Media receives more than 25 million visitors a month. National Geographic has funded more than 10,000 scientific research, conservation, and exploration projects and supports an education program promoting geography literacy. For more information, visit www.nationalgeographic.com.

For more information, please call 1-800-NGS LINE (647-5463) or write to the following address:

National Geographic Society
1145 17th Street NW
Washington, D.C. 20036-4688 U.S.A.

Your purchase supports our nonprofit work and makes you part of our global community. Thank you for sharing our belief in the power of science, exploration, and storytelling to change the world. To activate your member benefits, complete your free membership profile at natgeo.com/joinnow.

For information about special discounts for bulk purchases, please contact National Geographic Books Special Sales: ngspecsales@ngs.org

For rights or permissions inquiries, please contact National Geographic Books Subsidiary Rights: ngbookrights@ngs.org

ISBN 978-1-4262-1611-4

Printed in the United States of America

15/WOR/1